QUINE, NEW FOUNDATIONS, AND THE PHILOSOPHY OF SET THEORY

W. V. Quine's set theory, New Foundations, has often been treated as an anomaly in the history and philosophy of set theory. In this book, Sean Morris shows that it is, in fact, well motivated, emerging in a natural way from the early development of set theory. Morris introduces and explores the notion of set theory as explication: the view that there is no single correct axiomatization of set theory but rather various axiomatizations that all serve to explicate the notion of set and are judged largely according to pragmatic criteria. Morris also brings out the important interplay between New Foundations, Quine's philosophy of set theory, and his philosophy more generally. We see that Quine's early technical work in logic foreshadows his later famed naturalism, with his philosophy of set theory playing a crucial role in his primary philosophical project of clarifying our conceptual scheme and, specifically, its logical and mathematical components.

SEAN MORRIS is Associate Professor of Philosophy at the Metropolitan State University of Denver. He has written extensively on Quine and the history of analytic philosophy with a particular emphasis on Quine's work in the foundations of mathematics.

QUINE, NEW FOUNDATIONS, AND THE PHILOSOPHY OF SET THEORY

SEAN MORRIS
Metropolitan State University of Denver

CAMBRIDGE
UNIVERSITY PRESS

CAMBRIDGE
UNIVERSITY PRESS

University Printing House, Cambridge CB2 8BS, United Kingdom

One Liberty Plaza, 20th Floor, New York, NY 10006, USA

477 Williamstown Road, Port Melbourne, VIC 3207, Australia

314–321, 3rd Floor, Plot 3, Splendor Forum, Jasola District Centre, New Delhi – 110025, India

79 Anson Road, #06-04/06, Singapore 079906

Cambridge University Press is part of the University of Cambridge.

It furthers the University's mission by disseminating knowledge in the pursuit of education, learning, and research at the highest international levels of excellence.

www.cambridge.org
Information on this title: www.cambridge.org/9781107152502
DOI: 10.1017/9781316591321

First published 2018

Printed and bound in Great Britain by Clays Ltd, Elcograf S.p.A.

A catalogue record for this publication is available from the British Library.

Library of Congress Cataloging-in-Publication Data

Names: Morris, Sean (Professor of Philosophy), author.
Title: Quine, new foundations, and the philosophy of set theory / Sean Morris, Metropolitan State University of Denver.
Description: First edition. | New York : Cambridge University Press, [2019] | Includes bibliographical references.
Identifiers: LCCN 2018037322 | ISBN 9781107152502 (hardback)
Subjects: LCSH: Quine, W. V. (Willard Van Orman) | Set theory.
Classification: LCC B945.Q54 M67 2019 | DDC 160.92—dc23
LC record available at https://lccn.loc.gov/2018037322

ISBN 978-1-107-15250-2 Hardback

This book is dedicated to Bill Hart and to Haewon and John.

Contents

Preface

This book emerged out of my doctoral dissertation at the University of Illinois at Chicago, written some years ago under the direction of W. D. Hart. After completing the dissertation, I moved on to researching topics of a more general nature in Quine's philosophy, though always with an eye toward how his early work in logic and the foundations of mathematics shaped his general approach to philosophy. From time to time, I continued to present some of my views on Quine's philosophy of set theory and its contrasts with much contemporary philosophy of set theory. I remained surprised by – in discussions of the philosophy of set theory – how set theory was just about always and without question identified with Zermelo-Fraenkel set theory and its variants and the associated iterative conception of set. From this, it seemed to me that there might still be something to contribute to contemporary discussions of the philosophy of set theory by focusing on Quine's alternative approach – an approach that ignores the iterative conception, at least as a metaphysical view about sets, and that is a good deal more experimental and exploratory in nature, seeing set theory as a still largely unsettled area of the mathematical sciences.

In assembling the following list of acknowledgments, I could not help being reminded of the words of a recent Nobel laureate:

> Some are mathematicians
> Some are carpenters' wives ...
> We always did feel the same
> We just saw it from a different point of view.[1]

The support and inspiration I drew throughout the writing of this book, teachers, friends, and colleagues supported and inspired me in ways that were as varied as they were indispensable. The paths that many of these

[1] Bob Dylan, "Tangled Up in Blue," *Blood on the Tracks* (New York, NY: Columbia Records, 1972).

people chose are very different from mine, but without their support this project never would have been finished. First of all, I would like to thank my dissertation committee Juliet Floyd, W. D. Hart, Peter Hylton, Colin Klein, and Daniel Sutherland for their input into the dissertation that led to this book. W. D. Hart and Daniel Sutherland were especially helpful in continuing to urge me to rework the dissertation into a book, and Peter Hylton provided me with very thorough and detailed comments on the original dissertation. For encouragement and comments on portions of the book manuscript, I thank Gary Ebbs and Warren Goldfarb. I also thank Timothy Bays, Patricia Blanchette, Michael Detlefsen, and Curtis Franks for allowing me to present some of this material at the 2015 Midwest PhilMath Workshop. Dean Joan Foster provided me with research funds for this project by way of multiple Metropolitan State University of Denver (MSU) Liberal Arts and Sciences (LAS) mini-grants, and Rebecca Dobbin of the Dean's office helped me to obtain research materials. I also thank Roy Cook, Thomas Forster, Craig Fox, Matthew Makley, Nick Robertson (1980–2008), Marcus Rossberg, Paul Roth, Joseph Ullian, the staff of the Houghton Library at Harvard University, and my colleagues and the staff of the MSU Denver Department of Philosophy. I was also assisted greatly by comments on an early portion of the manuscript from two anonymous referees for Cambridge University Press. I also thank Hilary Gaskin at Cambridge University Press for her interest in this project as well as for her patience and understanding as I completed my manuscript.

Although I have mentioned him already, I want to single out Bill Hart, my dissertation advisor and friend. His confidence in and support of my work have always kept me working. I thank also his wife, Faith, for her support and encouragement. My friend Kimo Quaintance deserves a special mention for the many conversations we have shared over the years and for helping me to see connections where there seemed to be none. In addition, a late-night conversation with my friend Ben Nichols was crucial to reigniting my interest in this project.

Finally, I would like to thank my family, including Bill Burke, my mother, Mary, my father, Paul (1947–1997), and my brother, Dan. They have all, in various ways, helped me to complete this book. Most of all and most important, I thank my wife, Haewon, and our son, John. They have made this work possible and have given more to me than I ever could have asked for. Now that it is done, I hope we will all be spending a lot more time at the park together.

Introduction

Since the late 1960s, there has been widespread agreement among both philosophers and mathematicians that the only viable conception of a set is that of the iterative conception as embodied in Zermelo–Fraenkel (ZF) set theory and its related systems.[1] Indeed, in any discussion, set theory is now treated as practically synonymous with ZF set theory, and the iterative conception is treated as something near essential to the very nature of sets.[2] Other approaches to set theory are generally neglected if not dismissed outright in set-theoretic research. However, this has not always been the situation, as I hope to show in this book by examining W. V. Quine's set theory, New Foundations (NF), and its place in the philosophy of set theory more generally. I should make clear at the outset that what follows is not meant to be an argument in favor of NF over other set theories. Rather, the point will be that work in set theory generally, at least in its current state, should be conducted in the more pluralistic and pragmatic way that I take to be characteristic of what I will identify as the approach to set theory as explication. I describe the contrasting approach, which aims to discover a single correct notion of set, as set theory as conceptual analysis. Let me begin first, though, with some very general history that will be further developed in the main body of the text.

[1] The iterative conception of set had received some attention in the literature on the foundations of set theory, but its first thorough treatment came in George Boolos, "The Iterative Conception of Set," reprinted in his *Logic, Logic, Logic*, ed. Richard Jeffrey (Cambridge, MA: Harvard University Press, 1998), pp. 13–29.

[2] Admittedly, few if any philosophers of set theory talk of capturing the essence of a set; essence has become a fairly unfashionable notion in contemporary philosophy with its drive toward scientific methods and naturalism. Still, it seems to me that the near dogmatic adherence to the iterative conception suggests a commitment to it yielding something like the essence of a set. It is not hard to find examples of the view that sets *must* be like this, where the "this" is filled in by the iterative conception. We will see an example of the sort of view I have in mind in Chapter 5 in discussing Boolos's account of the iterative conception.

Around 1900, in light of the discovery of the set-theoretic paradoxes, the concept of set had to be rethought. Up to that point, this concept had been accepted as basically well understood, taken to be what philosophers and logicians had traditionally meant by the extension of a concept or of a predicate. The realization that not every predicate could determine an extension, on pain of paradox, left set theory at somewhat of a loss in accounting for its fundamental notion. The idea that a set is the extension of a predicate could be captured formally by what has become known as the comprehension principle, "$(\exists x)(\forall y)\ (x \in y \equiv \varphi)$," where x is not free in the formula "φ." But if we let "φ" be "$y \notin y$," for example, we can then very easily derive Bertrand Russell's paradox of the set of all non-self-membered sets. If this set is a member of itself, then it is not a member of itself; and if it is not a member of itself, then it is a member of itself – hence a contradiction.[3] This principle would have to be somehow restricted; but beyond the intuitive thought that a set is the extension of a predicate, which then also involves a commitment to sets to being extensional entities, the tradition did not have much more to go on.

Instead of looking for something inherent in the notion of a set that would yield an appropriately restricted existence principle, logicians and set theorists focused on what they took to be the mathematical value of the theory. In particular, set theory could provide a rigorous foundation for all of ordinary mathematics. The various number systems and operation on them, the notions of a relation and of a function, and even geometry could all be reduced to sets along with the sole relation of membership and logical vocabulary. In addition to this, set theory had to provide an account of the mathematics of the infinite, since Georg Cantor's discovery of an apparently consistent arithmetic of infinite numbers had been the real advance gained for mathematics by way of set theory. Previously, mathematicians had mostly thought any sort of actual infinite to be inherently contradictory. They had operated instead with only potential infinites. In its most basic form, this is the idea that we could always add 1 but that this yielded only another – though larger – finite number. Cantor's development of a rule-governed arithmetic of the infinite in a sense legitimized the infinite for the mathematical community. Hence a successful reworking of set theory was to restrict the comprehension principle enough to prevent the paradoxes but not so much that

[3] This is the simplest of the set-theoretic paradoxes, and it is perhaps for this reason that is the best known among them. There are other paradoxes, of course, which I will discuss further in the text. In particular, there are also Cantor's paradox of the largest cardinal number and the Burali-Forti paradox of the largest ordinal number.

it would lack the power it needed to capture ordinary mathematics as well as the mathematics of the infinite. Such broad criteria left much freedom for choosing how to do this. As we will see, the two main approaches that emerged were Ernst Zermelo's set theory and its related systems and Russell's theory of types. Both were considered to be legitimate approaches to set theory, and both were used widely in foundational studies for mathematics through at least the early 1960s; after that, Zermelo's theory rose to dominance.[4] Neither was considered to capture anything like the essence of a set unless the essence is minimally understood as including some restricted version of the comprehension principle, an axiom of extensionality (sets are identical when they have the same members; this is a distinguishing feature of sets, so much so that it would be hard to consider a theory that lacked it as set theory), and enough strength to capture what made set theory mathematically interesting in the first place.

This period of set theory up through the 1950s, then, was largely a pragmatic undertaking, involving much exploration and experimentation and a certain open-mindedness about how set theory should be done. In addition to the main systems of Russell and Zermelo, many variants of these systems were also studied.[5] There was much interest in whether these different set theories could decide certain fundamental questions about sets that were still open, such as the axiom of choice (AC) or Cantor's continuum hypothesis. There was also a general interest in the comparative study of the various options: How strong were they in comparison to each other, how much mathematics could they prove given their various assumptions, or how much could they tell us about the infinite? Aside from these main systems, there were other sets theories that diverged from them, though they, too, would adhere to the basic criteria

[4] It was dominant at least as a set theory. Russell's type theory tended to get relegated to logic as a higher-order logic. Although this is an important and interesting topic, I will not consider it further here, as it would take me well beyond my primary concern with the philosophy of set theory.

[5] Many of the standard texts of this period clearly reflect this approach to set theory. See, for example, the many set theories studied in A. A. Fraenkel, Y. Bar-Hillel, and A. Levy's *Foundations of Set Theory*, 2nd rev. edn (Amsterdam: North-Holland, 1973), or Mendelson's *Introduction to Mathematical Logic* (Boca Raton, FL: CRC Press, 1997). As we will see, Quine's *Set Theory and Its Logic*, rev. edn (Cambridge, MA: Harvard University Press, 1969), is certainly part of this tradition in set theory. This is quite different from most contemporary set theory texts, which focus pretty strictly on Zermelo's set theory and its variations. This may now be starting to change again in light of the growing interest and value found in the study of non-well-founded set theories. Some standard introductory set theory texts are starting to cover this topic; see, for example, Keith Devlin, *The Joy of Sets: Fundamentals of Contemporary Set Theory*, 2nd edn (New York, NY: Springer-Verlag, 1993), Ch. 7, and Karel Hrbacek and Thomas Jech, *Introduction to Set Theory*, 3rd rev. and expanded edn (New York, NY: Marcel Dekker, 1999), Ch. 14, Sec. 3.

that had been set for a viable set theory. Such is the set theory that is the main topic of this book: Quine's *New Foundations*.

Quine first put forward his NF in the 1937 paper for which this set theory is named.[6] It is strikingly different from most other set theories in that it allows for a universal set, that is, a set of all sets.[7] Such "big" sets as this one, along with the set of ordinal numbers and the set of cardinal numbers, have often been thought to be the source of the set-theoretic paradoxes of Russell, Cesare Burali-Forti, and Cantor, respectively, and so have generally not received widespread attention in the literature. But as we will see in the opening chapters, such sets were not always so easily dismissed or altogether ignored. Early practitioners did consider such sets – not always favorably, but they were certainly not far from mainstream thinking about set theory, whereas they typically are thus positioned these days. Questions about these big sets were a real concern, raising questions that needed to be answered if set theory was to go on. For example, if big sets were the source of the paradoxes, on what grounds could they be ruled out as illegitimate? Or if they were to be included as part of the set-theoretic universe, how could they exist without giving rise to contradiction? These were genuine and important issues to consider during set theory's founding and development by Cantor, Russell, and Zermelo. In the more recent development of set theory, these issues were not so much resolved as just simply faded from view as the two main set theories – ZF set theory and Russell's theory of types – emerged, both denying the existence of these big sets.[8] With this, we lost any mathematical investigation into what Cantor identified as the absolute infinite.[9] As I have already noted, this early period of set theory was not one of dogmatism, and research in set theory remained fairly pluralistic through the 1950s. With many important questions still open, chief among them being the continuum hypothesis, no one axiomatization of set theory was taken as definitive, and much research

[6] W. V. Quine, "New Foundations for Mathematical Logic," *American Mathematical Monthly* 44:2 (1937), pp. 70–80.

[7] As we will see later, it is not quite as deviant as it is often portrayed to be. While it certainly does behave in unusual ways, particularly when its "big" sets are under consideration, it also develops fairly naturally out of insights gathered from Russell and Zermelo. We will see that, in a way, it is a generalization of ideas that Quine took from both of them.

[8] In his early work, Russell himself thought it near obvious that these big sets must exist, and as we will see, it was this belief that brought him to the discovery of Cantor's paradox.

[9] Cantor himself did not think that the absolute infinite could be investigated mathematically, but for theological rather than mathematical reasons.

concerned comparative accounts of the various axiomatizations that were available.[10] It was in this context of pluralism and tolerance that Quine introduced NF.

This book consists of seven chapters, divided into three main parts, though all related in some way. The first part of the book is concerned primarily with the early development of set theory, including the emergence of Quine's NF. In Chapter 1, I present a brief history of set theory from its beginnings with Cantor through to the discovery of the paradoxes, and then, in Chapter 2, I turn to the main proposals offered by Zermelo and Russell for resolving them. In these first two chapters, I aim to bring out two important points that I mentioned previously. The first is that to the extent to which there was any single shared conception of a set at the founding of the theory, it was the idea that a set is the extension of a predicate. The second point is that after the paradoxes, the success of a set theory was to be judged by its ability to explain the infinite and to serve as a framework for reconstructing ordinary non-set-theoretic mathematics. This gives rise to the approach to the philosophy of set theory that I describe as set theory as explication. This is the idea that there is no uniquely correct notion of a set beyond its being the sort of object that fulfills the minimal criteria that I have mentioned already: It is an extensional entity, and it must somehow be restricted enough to avoid the paradoxes but not so much so that it ceases to be capable of fulfilling its intended role in mathematics. I take this notion of an explication from Quine, who himself took it, as we will see, from Rudolf Carnap and Russell. Quine himself states in his book *Word and Object* that for an explication, "[w]e fix on the particular functions of the unclear expression that make it worth troubling about, and then devise a substitute, clear and couched in terms to our liking, that fills those functions."[11] In this view, I argue, set theory is developed pragmatically in an experimental and exploratory spirit. This approach is to be contrasted with what I describe as set theory as conceptual analysis. This view takes it as a given that there is a single correct set theory and, through conceptually analyzing the notion of set itself, we can discover what it is. This latter view appears to be dominant in the contemporary philosophy of set theory, with its adherence to the iterative conception of set and the general unwillingness to entertain set theories that run contrary to this

10 Fraenkel and Bar-Hillel's *Foundations of Set Theory*, first published in 1958 with its last edition published in 1973, is a classic of this sort of work in set theory.

11 W. V. Quine, *Word and Object* (Cambridge, MA: MIT Press, 1960), pp. 258–9.

conception.[12] We will see examples of this attitude mostly in later chapters, where I discuss NF and its accompanying philosophy in the context of the iterative conception.

In Chapter 3, which concludes the opening line of thought with its historical focus on the paradoxes, I present Quine's NF as combining the insights of both Zermelo and Russell into resolving the paradoxes. Here, I emphasize that Quine's exploratory and experimental approach fits squarely within the tradition of explication. Thus NF is far from an anomaly in the history of set theory, though it is often treated as such within the context of contemporary philosophy of set theory. In a way, then, the current dominance of the iterative conception is much more the anomaly in the history of set theory.

The second part of the book comprises two chapters in which I focus more directly on Quine's philosophy of set theory within the context of his philosophy as a whole. The focus here is not so much on NF specifically as on the philosophical thinking that informed Quine's approach to set theory generally. In Chapter 4, I focus primarily on how Quine's philosophy of set theory developed out his early engagement with Russell's logicism, particularly as put forward in Russell's *Principles of Mathematics* and the more technical working out of this project in Whitehead and Russell's *Principia Mathematica*.[13] Whereas Chapter 2 emphasizes Russell's foundational work as part of the tradition of set theory as explication, Chapter 4 tells a broader story, emphasizing some of the tensions between Russell's mathematical and philosophical approaches to the foundations of mathematics. It is in the former where we find the idea of set theory as an explication. I argue that we can understand Quine's own particular conception of set theory as explication as emerging out his attempt to come to terms with the tensions in Russell's work. The final part of Chapter 4 returns to NF to emphasize the philosophy that accompanies it. In particular, we see in Quine an

[12] There is nothing in the iterative conception that demands that it be approached from the side of conceptual analysis. After all, as I have already remarked, the iterative conception is now most commonly taken to be embodied in the axioms of Zermelo's theory, which originally emerged out of the explicative approach. Indeed, I take it that there are good mathematical reasons to now favor a set theory such as ZF, but these are not the reasons that are typically offered in favor of the iterative conception over other conceptions of set. More typical are appeals to its rather intuitive character. This is an argument that I reject, as we will see throughout this book in various ways.

[13] Bertrand Russell, *Principles of Mathematics*, 2nd edn (London: George Allen and Unwin Ltd., 1937[first edition 1903]). Alfred North Whitehead and Bertrand Russell, *Principia Mathematica*. 3 vols. 2nd edn (Cambridge: Cambridge University Press, 1925, 1927 [first edition 1910, 1912, 1913]).

attitude that is very critical of what he identifies as the more metaphysical approaches to the philosophy of set theory.

In Chapter 5, I continue to focus on the philosophy that informs Quine's approach to set theory, but I turn to his more developed and more general approach to philosophy as a whole. The main theme here will be how Quine takes the simplification and clarification of our conceptual scheme as a main task for philosophy. His work in set theory is largely a contribution to this project, focusing on the mathematical portion of our conceptual scheme and reaching its culmination in *Set Theory and Its Logic*. Within the context of Chapter 5, I also discuss in some detail Quine's understanding of the notion of explication. Since my account of Quine often pushes him very close to Carnap, I conclude Chapter 5 by trying to locate where their differences lie, particularly with regard to their appeals to tolerance in their respective philosophies.

In Part III, I consider in more detail the other tradition in the philosophy of set theory, that of conceptual analysis, by examining in Chapter 6 some arguments in favor of the iterative conception of set, most notably championed by George Boolos. I argue that this preference ultimately rests on pragmatic considerations that support NF equally well. I then look at NF in more detail in this context to show that, like ZF-style set theories, it largely satisfies the criteria offered earlier for judging a set theory successful. That is, it appears to eliminate the paradoxes while also providing an account of the infinite and serving as a framework for reconstructing the rest of mathematics within it. In addition, I emphasize that contemporary set-theoretic research tends to focus on set theory itself as a subject for mathematical investigation and thus no longer limits itself to being merely a foundation for other branches of mathematics. NF, with its "big sets" such as the universal set, the set of cardinal numbers, and the set of ordinal numbers, is thus more consonant with current mathematical practice regarding set theory than might have been thought. It gives us a way to investigate the mathematical behavior of such sets, which ZF and related systems exclude. Again, my ultimate aim is not to argue for NF as the correct set theory over other set theories but to show that the concurrent development of various set theories furthers our understanding of the set-theoretic universe and so also our mathematical knowledge generally.

In Chapter 7, I examine some potential problems with NF. NF's development of arithmetic does not appear to proceed as elegantly as that of ZF. One way to deflect this challenge is to draw attention again to the fact that set theory has moved far beyond being merely a foundation for other branches of mathematics. I then also observe that one reason why ZF does give us an elegant account of, say, arithmetic is that ZF begins by ruling out certain troublesome sets from the start. In the more pragmatic view that I am putting forth, this a perfectly acceptable starting point. But I then argue that it would also be perfectly acceptable not to rule out these sets, particularly when our interest is in the mathematics of the set-theoretic universe as a whole. Within this context, I also consider the status of AC in NF. NF is notorious for disproving this axiom, which is used crucially throughout mathematics. However, I show that – much as in the case of arithmetic – this apparent failing arises only because NF includes sets that ZF does not. In NF, AC fails only in full generality. Appropriately restricted, enough of AC is present to satisfy the usual mathematical needs. Again, I show that such a failing of AC is hardly out of line with the set-theoretic tradition. For much of its history, AC has been treated carefully by set theorists, often flagging where it is used in proofs or in trying to re-prove results so that they do not rely on the axiom. Although much of contemporary set theory has come to treat AC as constitutive of the behavior of sets, even considering the principle obvious, this seems unfounded. After all, AC is well known for its highly unintuitive consequences, such as the Banach–Tarski paradox. In consequence, the truth of AC seems far from settled, thus leaving as a worthwhile endeavor the investigation of set theories for which it fails.

My aims in the book are broadly philosophical in that, most generally, I want to urge that logic and mathematics continue to be fruitful sources for thinking about philosophical problems that have been with us since philosophy's beginning. In particular, this book raises again the age-old question of how we come to acquire our nonempirical knowledge. In arguing that set theory is pragmatically developed, I want to suggest that mathematics in its foundations develops in ways that share much with the natural sciences and therefore is not necessarily the paradigm of a priori knowledge that philosophers have often thought it to be.

Before concluding this introduction, I should comment further on what I have been describing as an approach to set theory as explication. I have emphasized that this approach to set theory is pragmatic. Some of my exposition up to this point and in what follows may suggest that the pragmatic approach to set theory is unique to my account of set theory as explication and that it no longer has much hold on contemporary philosophy of set theory. This is not entirely true. A number of commentators have noted that this tradition has a firm place in the history of set theory.[14] Furthermore, a pragmatic approach to set theory is not entirely absent from contemporary set theory.[15] But in accord with the account that I have been laying out, this pragmatism happens only within ZF-style set theories. It does not extend to alternative axiomatizations. This is just the sort of dogmatism in set theory that I will question in what follows. With this in mind, let me now present Quine's NF and its place in the philosophy of set theory.

[14] For some examples of the pragmatic tradition in the history of set theory, see Gregory H. Moore, "The Origins of Zermelo's Axiomatization of Set Theory," *Journal of Philosophical Logic* 7:1 (1978), pp. 324–5; M. D. Potter, "Iterative Set Theory," *The Philosophical Quarterly* 43:171 (1993), p. 181; and Heinz-Dieter Ebbinhaus, *Ernst Zermelo: An Approach to His Life and Work* (New York, NY: Springer, 2007), p. 80. I should also add that I have previously described what I am now calling set theory as explication as the pragmatic approach to set theory. However, this label seemed to cause too much confusion with American pragmatism as a general philosophical position. For suggesting the new terminology, distinguishing between set theory as explication and set theory as conceptual analysis, I thank Gary Ebbs.

[15] We find this in Penelope Maddy's recent *Defending the Axioms: On the Philosophical Foundations of Set Theory* (New York, NY: Oxford University Press, 2011), in her discussions of intrinsic versus extrinsic justifications of the axioms of set theory. The former lines up with conceptual analysis and the latter with explication.

PART I

Set Theory's Beginnings

CHAPTER 1

Cantor and the Early Development of Set Theory

This opening chapter provides a basic account of the development of set theory up to about 1900 and the discovery of the set-theoretic paradoxes. Here, I focus specifically on the most important results of Georg Cantor (1845–1918), the founder of set theory. Two important themes begin to emerge here that will receive greater attention in later chapters. First, we will see that Cantor developed set theory with the aim of making mathematical sense of the infinite. This comes in two intertwined parts: Cantor's attempts to legitimize the actual, or proper, infinite and his discovery of countable and uncountable infinities. Of particular importance is Cantor's effort to legitimize the infinite as actually existing by providing mathematical rather than philosophical arguments. This will be especially important for my discussion in Chapter 2 of how the notion of set was eventually constrained to consistency after the paradoxes. The second theme, which is most important for my overall aims, is that Cantor developed set theory as a *mathematical* theory, addressing concerns within mathematics and not about some prior philosophical conception of a set.

This will continue to be a major theme throughout this book and one that I will develop most directly in Chapter 4 in considering W. V. Quine's particular view of the philosophy of set theory. Briefly put for now, I describe this tradition in the philosophy of set theory as set theory as explication. This will be contrasted with what I will later describe as set theory as conceptual analysis. This latter tradition is now much more prominent in the philosophy of set theory and can be roughly characterized by the view that there is a uniquely correct conception of set theory and we will be able to discover it by way of conceptual analysis of the notion of set. Contemporary philosophers of set theory have largely followed this path and are now in general agreement that the iterative conception as embodied in the axioms of Zermelo–Fraenkel set theory is what is meant by set theory. The other tradition – set theory as explication – has been largely forgotten, if it has ever really been accorded its place as a distinct tradition at all. Quine is perhaps the most prominent adherent of this view. Rather than aiming at a single correct theory of set, it attempts to identify what made set theory worthwhile in the first place and develops a set theory that is in accord with these features. In doing so, it allows for a plurality of set theories and is exploratory and experimental in nature. It does not, however, rule out our eventually settling on a single set theory. However, doing so would be largely the result of certain pragmatic concerns and not because we had somehow arrived at, say, the essence of sethood. We will have to wait until Chapter 4 and beyond to fully develop these two traditions and examine their relative philosophical merits. Let us instead consider set theory at its very beginnings.

1.1 The Beginnings of Set Theory as a Mathematical Discipline

Set theory began as a mathematical discipline – though often deeply connected to the philosophical views of its founders – during the nineteenth century as the calculus was being treated to greater rigorization. During this period, troubles began to arise around the unrestricted use of functions and the related notions of convergence and continuity.[1] Since the beginnings of the calculus with Gottfried Leibniz and Isaac Newton, the notion of a function had been gradually expanded from analytic expressions to arbitrary correspondences. By the late eighteenth century, however, Leonhard Euler, along with Joseph-Louis Lagrange, had made the first significant expansions of the concept of a function by introducing methods relating to infinite series. Before their work, mathematicians had held to the view that a function must always have the same analytic expression regardless of its domain. The expansion of this concept allowed for a function to be expressed differently depending upon its domain. As the use and variety of what was to count as a function continued, mathematicians gradually adopted a broader understanding of the concept of a function itself. But with this came new difficulties regarding the properties of functions, chief among them what was meant by continuity, differentiability, and integrability. Furthermore, many of these concepts had no more than an intuitive foundation expressed in

[1] Akihiro Kanamori, "The Mathematical Development of Set Theory from Cantor to Cohen," *The Bulletin of Symbolic Logic* 2:1 (March 1996), pp. 1–2. I follow Morris Kline, *Mathematical Thought from Ancient to Modern Times* (New York, NY: Oxford University Press, 1972), pp. 947–77 for the account of the rigorization of the calculus that follows over the next few pages.

temporal, dynamical, or geometric terms. Mathematicians of what has become known as "the critical movement" began urging that a foundation for analysis be constructed more rigorously and from arithmetical concepts. While geometry had frequently been used in the past for such foundational work, the rise of non-Euclidean geometries in the nineteenth century brought into question the privileged status that geometry seemed to hold. If any branch of mathematics remained primary, it was generally thought to be arithmetic. Central among those urging analysis in this direction were Bernard Bolzano, Augustin-Louis Cauchy, Gustav Dirichlet, and Karl Weierstrass.

By the second half of the eighteenth century, Euler and Lagrange had taken a function to be continuous when the same expression held and discontinuous at points where the expression changed. However, this was not the modern understanding of these notions or of a function generally. Dirichlet gave the modern definition of a function in 1829, stating that y is a function of x when, for each value of x in a given interval, there is exactly one value of y. In line with mathematicians' acceptance of an increasingly broad conception of a function, he added that within the given interval y could depend upon x according to more than one law and that the dependence did not have to be expressed by way of mathematical operations. With the concept of a function so refined, a proper understanding of continuity and discontinuity also emerged, primarily as the result of work by Bolzano and Cauchy. Bolzano defined a function $f(x)$ as continuous in an interval if at any x in the interval, the difference $f(x + \omega) - f(x)$ can be made as small as one wishes by taking ω sufficiently small. Notable in his definition is the elimination of infinitesimals – infinitely small but nonzero quantities – used throughout the eighteenth century but with no rigorous basis.[2] Cauchy similarly defined the notion of continuity while denying infinitesimals. A key innovation here was his introducing the notion of a limit. He replaced talk of infinitesimals with talk of limits describing a variable quantity as becoming infinitely small when its numerical value decreases indefinitely so as to converge to a limit zero. He identified these variable quantities as infinitesimals but without the metaphysical overtones associated with their use in Leibniz and his followers. Weierstrass contributed further to rigorizing the calculus by removing talk of a variable approaching a limit.

[2] Infinitesimals returned with a rigorous basis in the twentieth century in the work of Abraham Robinson, first published in his "Non-Standard Analysis," reprinted in his *Selected Papers*, vol. 2, eds. H. J. Keisler et al. (New Haven, CT: Yale University Press, 1979), pp. 3–11.

He stressed instead that the variable was just a letter standing for any one of a set of values that may be assigned to the variable. In this way, he eliminated temporal and dynamic intuitions from the rigorous formulation of the calculus.

The derivative, too, stood in need of a clear definition. Mathematicians before the nineteenth century appealed to geometric intuitions and the mysterious notion of an infinitesimal. Again, Bolzano gave the first clear account in 1817, defining the derivative of $f(x)$ as the quantity $f(x)$, which the ratio $[f(x + \Delta x) - f(x)]/\Delta x$ approaches indefinitely close as Δx approaches zero through positive and negative values. Cauchy added to this definition by unifying it with Leibnizian differentials, defining *dx* as any finite quantity and *dy* as the derivative $f(x)dx$ so that the differentials are accounted for in terms of derivatives. Finally, he clarified the relationship between $\Delta x/\Delta y$ and $f(x)$ by way of the mean value theorem: $\Delta y = f(x + \theta\Delta x)$, where $0 < \theta < 1$. While the work of Bolzano and Cauchy yielded rigorous accounts of continuity and the derivative, this still left questions about continuous functions and differentiability. In particular, it had previously been thought that all continuous functions must be differentiable, but in 1834, Bolzano found an example of a continuous function with no finite derivative at any point, though he did not publish it. However, Bernhard Riemann did publish an example of such a function in 1868, forcing the mathematical community to recognize that continuity and differentiability had to be distinguished. Such counterexamples to previous thinking on these topics led mathematicians to be all the more skeptical of appeals to intuition and geometrical approaches to the calculus.

These developments led also to concerns over the integral. Newton had shown how the integral could be found by simply reversing differentiation, which is still the standard approach. But now questions arose about the integral for discontinuous functions. Leibniz had already examined calculating areas and volumes in terms of a sum of elements such as rectangles and cylinders. Eighteenth-century mathematicians had followed him in this, but again with the usual lack of rigor. Cauchy began to correct this. Joseph Fourier had already shown that the Leibnizian view worked to calculate the integral for discontinuous functions, but this still left the problem of expressing the integral analytically for discontinuous functions. Cauchy took up this problem, emphasizing that that integral should be given as the limit of a sum rather than in the Newtonian way as the inverse of differentiation. His aim, achieved in 1823, was to show that the integral could be given arithmetic expression.

Ultimately, however, his methods proved too restrictive to apply to the many irregular functions that had entered into analysis with the broadening of the concept of a function. Riemann stepped in to correct this in an 1854 paper on the trigonometric series in which he sufficiently generalized the notion of an integral, using an example of an integrable function with an infinite number of discontinuities in every arbitrarily small interval to show the success of his definition. Some outstanding issues remained, but by 1875, the integral had received sufficiently rigorous and general expression.

We have seen here how Bolzano, Cauchy, Dirichlet, and Weierstrass, among others, addressed the various foundational problems in the calculus by further articulating such concepts as convergence and continuity and by replacing infinitesimals with the notion of a limit in terms of the now familiar epsilon-delta language. The result of their work not only eliminated the apparent flaws in the calculus but, more important for our purposes, also restored an emphasis on foundational work in mathematics and on deductive rigor in particular. Making sense of these new functions in terms of an infinite series was achieved only through their careful specification by means of deductive methods. Cantor emerged from this tradition that restored proof as the focal point of mathematics, thus furthering the greater abstraction and generality characteristic of modern foundational work in mathematics.[3] Indeed, Cantor also made direct contributions to the rigorization of the calculus, and his founding of set theory emerged directly out of this research. Perhaps his most fundamental contribution was an account of the real numbers themselves. For most of the developments toward rigorization that we have seen so far, the structure of the real number line was assumed to be well enough understood. But as we have seen, greater clarification of one concept often exposed a lack of clarity in another. This was also the case for the real numbers themselves. For example, Cauchy was unable to prove his criterion for the convergence of sequence sufficient specifically because the structure of the real numbers themselves had not yet been subjected to rigorization. It was thought that an intuitive understanding was enough.[4] We will return to this topic soon enough, but let us begin with Cantor's earliest mathematical contributions to the emergence of set theory.

Cantor's first important move toward set theory came in 1870 in proving a uniqueness theorem for trigonometric series: If a trigonometric

[3] Kanamori, "The Mathematical Development of Set Theory from Cantor to Cohen," p. 2.
[4] Kline, *Mathematical Thought*, p. 979.

series converges everywhere to zero, then all of its coefficients are equal to zero.[5] He generalized this result in his "On the Extension of a Theorem of the Theory of the Trigonometric Series" [1872][6] to obtain the result: For a collection of real numbers P, let P' be the collection of limit points for P, and let $P^{(n)}$ be the result iterating this operation n times. If a trigonometric series converges everywhere to zero except on a P where for some n, $P^{(n)}$ is empty, then all of its coefficients are equal to zero. While mathematically significant in itself, Cantor's result had important consequences for the beginnings of set theory. His key move here was in considering collections of real numbers specified by an operation. Here, Cantor moves beyond considering just individual limit points to considering collections of them. In specifying such collections by an operation (later, by a law), Cantor was not far from the idea that sets are the extensions of concepts, which located the notion of a set as the extension of a concept deep within the logical tradition.[7] This was, of course, as we will see, the notion of a set that was to yield the paradoxes but was to be maintained in the post-paradox world in a sufficiently restricted sense – restricted enough to restore consistency yet preserve all that was mathematically valuable in the theory.

Just a couple of years later, Cantor made another important move toward the full development of modern set theory: He showed how to build up the real numbers from collections of rational numbers.[8] While Richard Dedekind's cuts are perhaps now the more familiar construction of the real numbers from rational numbers, Cantor construed the real numbers as what he called "fundamental sequences" of rational numbers, or what are now often called Cauchy sequences.[9] Elaborating on the motivations for this work some years later in his 1883 *Grundlagen*,

[5] Georg Cantor, "Über einen die trigonometrischen Reihen betreffen Lehrsatz," in Cantor, *Gesammelte Abhandlungen: Mathematische und Philosophischen Inhalts*, ed. Ernst Zermelo (Hildesheim: Georg Olms Verlagsbuchhandlung, 1962), pp. 71–9.

[6] Cantor, *Gesammelte Abhandlungen*, pp. 92–102.

[7] Kanamori, "The Mathematical Development of Set Theory from Cantor to Cohen," p. 2. On the relationship between sets as the extensions of concepts and the specification of a set by an operations, see José Ferreirós, *Labyrinth of Thought: A History of Set Theory and Its Role in Modern Mathematics* (Boston, MA: Birkhäuser Verlag, 1999), pp. 47–53, 143. See also George Boolos, "The Iterative Conception of Set," in *Logic, Logic, and Logic*, ed. Richard Jeffrey (Cambridge, MA: Harvard University Press, 1998), pp. 13–14. I will return to this point in more detail in later chapters.

[8] Cantor, *Gesammelte Abhandlungen*, pp. 92–6.

[9] The fundamental sequences are defined by the same property Cauchy used for his criterion of convergence; see Ferreirós, *Labyrinth of Thought*, p. 128.

Cantor singles out three definitions of the real numbers: Weierstrass's, Dedekind's, and his own. Remarking on Weierstrass's definition, Cantor observes:

> One sees here that the creative element which binds the set with the number defined through it lies in the *formation of sums*; but it must be emphasized as *essential* that only the summation of an always *finite* Anzahl of rational elements is used and that the number b to be defined is *not* set at the beginning as equal to the sum Σa_ν of the infinite series (a_ν); this would be a *logical error*, because the definition of the sum Σa_ν is only reached by equating it with the *finished* number b which is necessarily defined earlier. I believe that this logical error, which was first avoided by Weierstrass, was in earlier times the universal practice, and was not noticed because it belongs to those rare cases in which actual errors can do no significant harm to calculations.—Nevertheless I am convinced that all the difficulties which have been found in the concept of the irrational are connected with the indicated error, whereas if we avoid this error the irrational numbers take root in our minds with the same precision, distinctness, and clarity as the rational numbers [Cantor's italics].[10]

Of particular interest here is the "logical error" that Cantor describes, which will be a theme prominent in the development of set theory, as we will see in following chapters. The basic idea in Cantor's logically correct approach is that we do not first assume the existence of a rational number and then try to find some set of objects that will serve as its definition. Rather, we assume only the antecedently accepted objects – in this case, the rational numbers – and define the new objects, the real numbers, in terms of them. There is not, then, a preexisting set of objects, the real numbers, of which we are trying to get the single correct account. Instead, we identify the key mathematical properties of the real numbers – those properties that make the real numbers worthwhile – and then try to construct them out of antecedently well-understood objects.[11] Indeed, Cantor goes on to show that all three of the proposed definitions do have the mathematical properties of the real numbers. By doing so, we bestow the real numbers "with the same precision, distinctness, and clarity as the rational numbers." Furthermore, the three definitions are to be distinguished only in terms of their utility (and Cantor preferred his own definition on these grounds); otherwise, they all have

[10] Cantor, *Foundations*, sec. 9, para. 4. I have used Ewald's translation and so have left "Anzahl" untranslated. While there is some dispute over how to best translate this term, it can be basically understood as "ordinal number."

[11] As we will see, this is very much the attitude that Quine adopts toward the notion of explication.

equal right to the claim of being the real numbers.[12] Kanamori notes three particularly important aspects of Cantor's construction of the real numbers for his further developing of set theory. The construction in terms of fundamental sequences led him to focus more explicitly on infinite collections as such, to view them as unitary objects, and finally, to allow for arbitrary possibilities of such objects, which connects back to Cantor's idea of specifying sets by an operation or rule.[13]

We should note two further things about Cantor's defining of the real numbers that are particularly important to my own aims. One is his emphasis on construction so as to avoid the "logical error" of previous definitions. This approach to the foundations of mathematics becomes especially prominent in Bertrand Russell's work, with his emphasis on logical construction. The other, which is even more important for my overall aims, arises in Cantor's comparison of the three different definitions. As long as they all yield the mathematical properties required of the real numbers, there is no distinguishing among them except perhaps on pragmatic grounds, such as ease of use, elegance, or simplicity. As we will see in Chapter 2, in light of the paradoxes, something like this attitude becomes standard as an approach to set theory itself.

It was Cantor's next move – to prove that the real numbers are uncountable – that led to his full-blown development of transfinite set theory. As Kanamori vividly makes the point, "[s]et theory was born on that December 1873 day when Cantor established that the collection of real numbers is uncountable, and in the next decades the subject was to blossom through the prodigious progress made by him in the theory of ordinal and cardinal numbers."[14] During this period, Cantor began considering infinite iterations of his operator P', where

$$P^{(\infty)} = \cap_n{}^{\infty} P^{(n)}, P^{(\infty+1)} = P^{(\infty)'}, P^{(\infty \cdot 2)}, \ldots, P^{(\infty^2)}, \ldots,$$
$$P^{(\infty^\infty)}, \ldots, P^{(\infty^{\infty^\infty})}, \ldots$$

He also began to investigate infinite collections and real numbers and infinite enumerations as such. Combined, these moves led him to the basic concepts used in the study of the continuum and to the

[12] Cantor concludes his discussion with just such a comparison of the three definitions and explains why he prefers his own (paras. 19–23). On this point, see also Joseph Warren Dauben, *Georg Cantor: His Mathematics and Philosophy of the Infinite* (Princeton, NJ: Princeton University Press, 1979), p. 37.

[13] Kanamori, "The Mathematical Development of Set Theory from Cantor to Cohen," p. 3.

[14] Ibid.

formulation of the transfinite numbers.[15] His first major result in this direction was to prove that the set of real numbers is uncountable. The proof first appeared in print in Cantor's 1874 article "On a Property of the Totality of All Real Algebraic Numbers."[16] He first establishes that the algebraic numbers are countable, where a real number ω is algebraic if there exists a positive integer n, and integers $a_0, a_1, \ldots, a_n, a_n \neq 0$, such that

$$a_n\omega^n + a_{n-1}\omega^{n-1} + \cdots + a_1\omega + a_0 = 0.$$

He then proceeds by reductio to show that for any countable sequence of real numbers, every interval contains a real number that is not in the sequence:

Let the set of real numbers be countable. Then each real number ω can be sequenced by indexing them with natural numbers n:

$$\omega_1, \omega_2, \omega_3, \ldots, \omega_n, \ldots$$

Now, given an interval (α, β) that is a subset of real numbers, it is possible to find at least one real number η such that η fails to be listed as an element of the sequence. To find such a number η, let $\alpha < \beta$ and pick the first two numbers α', β' of the preceding sequence in the interval (α, β). These form another interval (α', β'). Then continue this procedure to yield a sequence of nested intervals through (α^n, β^n), where α^n, β^n are the first two numbers of the sequence in the interval $(\alpha^{n-1} \ldots \beta^{n-1})$. There are two possibilities to consider. If the number of constructed intervals is finite, then at most only one additional element of the sequence could lie in the interval (α^n, β^n). By choosing any real number η in the interval (α^n, β^n) not equal to the possible number from the sequence, we find a real number that is not listed in the sequence. On the other hand, if the number of constructed intervals is not finite, since the sequence $\alpha, \alpha', \ldots, \alpha^n, \ldots$ is bounded in (α, β), it has an upper limit α^∞, and similarly, the sequence $\beta, \beta', \ldots, \beta^n, \ldots$ has a lower limit β^∞. There are two further cases to consider. If $\alpha^\infty < \beta^\infty$, then as in the finite case, any real number η in the interval $(\alpha^\infty, \beta^\infty)$ would be the real number not listed in the preceding sequence. However, if $\alpha^\infty = \beta^\infty$, then $\eta = \alpha^\infty = \beta^\infty$, and η could not be listed as an element of the preceding sequence. Let $\eta = \omega_\rho$. For n a

[15] Ibid.

[16] Georg Cantor, "On a Property of the Totality of All Real Algebraic Numbers," in *From Kant to Hilbert: A Source Book in the Foundations of Mathematics*, vol. II, ed. William Ewald (Oxford: Clarendon Press, 1996), pp. 840–3.

sufficiently large index, ω_ρ would be excluded from all the intervals nested within (α^n, β^n). But by virtue of the construction, η must lie within (α^n, β^n). This is a contradiction, and hence the collection of real numbers is uncountable.[17] With this proof, Cantor secured the importance of the infinite as a legitimate area of mathematical research.

1.2 From the Potential to the Actual Infinite

In his next publication, "A Contribution to the Theory of Manifolds" [1878], Cantor began to focus more directly on bijective mappings between sets.[18] He defined two sets as having the same power if and only if there is a bijective mapping between them. Whereas his earlier result on the uncountability of the real numbers showed where such a mapping failed to hold, he now looked to see where these mappings did hold. It was these investigations that led him to fully develop the mathematics of the transfinite.[19] At the end of his investigations, however, he had been able to find only two powers of infinite sets and conjectured that every set of real numbers has the first power, that is, is countable, or has the power of the continuum. Cantor's attempts to solve this early version of the continuum hypothesis pushed his work forward, leading him to a more arithmetical approach in studying sets of real numbers as well to questions of set existence as such. At least where Cantor is concerned, Kanamori observes that "[s]et theory had its beginnings not as some abstract foundation for mathematics but rather as a setting for the articulation and solution of the *Continuum Problem*: to determine whether there are more than two powers embedded in the continuum."[20] We will see this theme continue into Chapter 2 in the attempts to come to terms with the paradoxes. We will see there that for Cantor, Russell, and Ernst Zermelo, what was important in regaining consistency was to maintain all that was mathematically valuable to the theory. Beyond this, there was no prior notion of set to be maintained.

Cantor's next major work – one of the most important in presenting his mature view of set theory – was "Foundations of a General Theory of Manifolds: A Mathematico-Philosophical Investigation into the Theory

[17] I have followed Dauben. *Georg Cantor*, pp. 51–2, in presenting Cantor's proof, but Kanamori, "The Mathematical Development of Set Theory from Cantor to Cohen," p. 6, also sketches it, although in less detail.

[18] Cantor, *Gesammelte Abhandlungen*, pp. 119–33.

[19] Kanamori, "The Mathematical Development of Set Theory from Cantor to Cohen," pp. 4–5.

[20] Ibid., p. 5.

of the Infinite" [1883], the *Grundlagen*. In this work, he introduces his theory of ordinal numbers and the notion of well-ordering.[21] He shifts focus away from the infinitely indexed operator P^{ν} used in his paper on the trigonometric series and turns his attention to the indexes themselves, that is, to what become his ordinal numbers. He also makes a notable notational change in moving from the symbol "∞," which traditionally indicates the potential infinite, to "ω," the final letter of the Greek alphabet, to represent the infinite as a completed whole, that is, the actual infinite. In another terminological shift, indicating his turning away from subsets of real numbers and to abstract sets themselves, Cantor stops speaking of point manifolds and instead talks just of sets. The major achievement of the *Grundlagen*, as these changes indicate, was to single out the transfinite numbers as both an autonomous and systematic extension of the finite numbers. Without this move, Cantor would have had no way to progress beyond the finite, and research in abstract set theory and on the continuum would have come to a standstill.[22]

In contrast with Cantor's new approach, mathematicians had traditionally treated the infinite as a variable increasing beyond all limits or decreasing to arbitrary smallness. In being arbitrarily large or small, however, the infinite was only potential, with the finite remaining the fundamental notion. The basic idea was that there was always some finite number, large or small enough, available for the task at hand. Actual, or proper, infinities, that is, infinities considered as complete in themselves, were thought to be incoherent. Cantor aimed to show in the *Grundlagen* that this was not the case. A succession of actually infinite numbers could be developed with identifiable and determinate number-theoretic properties, making these transfinite numbers just as legitimate as the systems of finite numbers.[23] Before this work, Cantor had no simple definition of the powers beyond the denumerable infinite. The least power among the infinite sets was the countable set of natural numbers. His new transfinite numbers would allow for a natural definition of powers beyond this countable infinity. To this end, he observed that the natural numbers resulted from the repeated addition of units to yield the sequence 1, 2, 3,... He calls this *the first principle of generation*, the successive generation of finite ordinal numbers by successive addition. This class, the number

[21] Georg Cantor, "Foundations of a General Theory of Manifolds: A Mathematico-Philosophical Investigation into the Theory of the Infinite," in *From Kant to Hilbert*, ed. Ewald, pp. 878–920.

[22] Kanamori, "The Mathematical Development of Set Theory from Cantor to Cohen," p. 5; Dauben, *Cantor*, p. 96.

[23] Dauben, *Georg Cantor*, p. 96.

class (I), had no largest element, but there was nothing to stop Cantor from introducing a new number ω expressing the natural order of the entire set (I). This number is the first number following the entire sequence of finite numbers in the set of natural numbers; it is the first transfinite number. But there was also no reason not to apply the first principle of generation again to yield additional transfinite numbers ω, $\omega + 1$, $\omega + 2, \ldots, \omega + n, \ldots$ Again, this sequence had no largest element, but Cantor then introduced another new number, 2ω, to represent this entire sequence in its natural order. He again applied the first principle of generation to yield the new sequence $2\omega, 2\omega + 1, 2\omega + 2, \ldots, 2\omega + n, \ldots$

This process then led Cantor to characterize what he called *the second principle of generation*. He explained that ω could be thought of as a limit approached by the sequence of natural numbers but never reached by it in the sense that ω is the first whole number after all the finite numbers in the set of natural numbers. So his second principle of generation states that if a sequence of numbers has no greatest element, then a new transfinite number can always be generated as the least number greater than all those in the sequence. So the successive application of the two principles of generation always allows for the possibility of generating a new number in succession to those previously generated numbers. Once Cantor had this second principle, he was able to define his second number class (II) as the collection of all numbers α formed from the two generating principles in a definite increasing succession in which all numbers proceeding α constitute a set with the same power as the first number class (I). He also indicated a third principle, *the principle of limitation*, that was to allow him to proceed to even higher classes of numbers, though he did not do much to develop this idea.[24]

In summarizing the significance of this work on the first and second number classes, Dauben notes the important difference mentioned earlier between the transfinite numbers and Cantor's earlier introduction of infinite symbols. Previously, Cantor had focused on derived sets of the second species and treated the transfinite symbols attached to his operator P^{v} as mere indices to identify and distinguish among the derived sets themselves. However, the transfinite numbers were themselves independent numbers, and this was necessarily so. Cantor aimed to use these very numbers in

[24] Dauben, *Georg Cantor*, pp. 97–8. Cantor presents his three principles of generation in sections 11 and 12 of the *Foundations*. We will see in Chapter 2 how Cantor avoided the paradoxes, the Burali-Forti paradox being the relevant one in this case. This approach to set theory is developed further by Michael Hallett in his *Cantorian Set Theory and Limitation of Size* (Oxford: Clarendon Press, 1984).

his further investigations into the powers of sets and the continuum. Hence he could not define the transfinite numbers in terms of the very sets he wished to study. They required an independent formulation so that they could be applied to the study of point sets and their powers. The transfinite numbers were to be understood as having an independent claim on reality equal to that of any of the other numbers.[25]

Another crucial aspect of the development of set theory introduced in the *Grundlagen* was the well-ordering principle, which Cantor described as "the law of thought that says that it is always possible to bring any *well-defined* set into the *form* of a *well-ordered* set – a law that seems to me fundamental and momentous and quite astonishing by reason of its general validity."[26] As Kanamori points out, this principle can be understood as part of the unity Cantor saw between the finite and transfinite numbers; just as the finite numbers can be well ordered, so can the transfinite numbers.[27] Cantor introduces the notion of a well-ordered set early in the *Grundlagen* stating that

> [a] well-ordered set is a well-defined set in which the elements are bound to one another by a determinate given succession such that (i) there is a first element of the set; (ii) every single element (provided it is not the last in the succession) is followed by another determinate element; and (iii) for any desired finite or infinite set of elements there exists a determinate element which is their immediate successor in the succession (unless there is absolutely nothing in the succession following all of them).[28]

These well-ordered sets were essential to his investigations into the transfinite numbers and in particular, in distinguishing finite from infinite sets. Here, he introduced the important concept of a numbering [Anzahl], where a numbering expressed the ordering of the elements of a given set. Later, he would identify these numberings as the ordinal numbers themselves. In drawing this connection between well-ordered sets and their numberings, Cantor was also able to further the transfinite numbers' claim to having objective reality. This objective reality of the transfinite numbers came from the existence of a well-ordered set whose order could be expressed by associating them with a number from the various transfinite number classes. To this end, Cantor aimed to show that for any countably infinite well-ordered set, there was always a

[25] Dauben, *Georg Cantor*, pp. 98–9.
[26] Cantor, "Foundations of a General Theory of Manifolds," p. 886.
[27] Kanamori, "The Mathematical Development of Set Theory from Cantor to Cohen," p. 6.
[28] Cantor, Foundations of a General Theory of Manifolds, p. 884.

number of the second number class (II) that uniquely represented its ordering.[29]

To illustrate this, Cantor considered the denumerable set (α_ν), which can be well ordered, for example, in any of the following ways:

$$(\alpha_1, \alpha_2, \ldots, \alpha_v, \alpha_{v+1}, \ldots)$$
$$(\alpha_2, \alpha_3, \ldots, \alpha_{v+1}, \alpha_{v+2}, \ldots, \alpha_1)$$
$$(\alpha_3, \alpha_4, \ldots, \alpha_{v+2}, \alpha_{v+3}, \ldots, \alpha_1, \alpha_2)$$
$$(\alpha_1, \alpha_3, \ldots, \alpha_2, \alpha_4, \ldots).$$

Cantor states that two well-ordered sets have the same numbering, or are similar, if they can be put into a one-to-one correspondence in such a way that preserves their respective orderings. So if α_n comes before α_m in the ordering of one set, then in the other set, the respective corresponding elements α'_n and α'_m must be ordered so that α'_n comes before α'_m. Such correspondences are always uniquely determined. So Cantor observed that given any two well-ordered sets, he could use the succession of natural numbers plus the transfinite numbers to identify similar well-ordered sets. Given any number α of the first or second number class, when taken together with all of its preceding elements, the numbering of all similar well-ordered sets is given by α uniquely. For example,

$$(\alpha_1, \alpha_2, \alpha_3, \ldots, \alpha_v, \alpha_{v+1}, \ldots)$$
$$(\alpha_2, \alpha_1, \alpha_4, \ldots, \alpha_{v+1}, \alpha_v, \ldots)$$
$$(1, 2, 3, \ldots, v, \ldots)$$

all have the same numbering ω. Similarly,

$$(\alpha_2, \alpha_3, \ldots, \alpha_v, \ldots, \alpha_1)$$
$$(\alpha_3, \alpha_4, \ldots, \alpha_{v+1}, \ldots, \alpha_1, \alpha_2)$$
$$(\alpha_1, \alpha_3, \ldots, ; \alpha_2, \alpha_4, \ldots)$$

have the distinct numberings $\omega + 1$, $\omega + 2$, and 2ω, respectively.[30]

Cantor also employed his distinction between number and numbering to bring out differences between finite and infinite sets. For finite sets, their numbering remains the same regardless of their ordering, whereas for infinite sets, there are sets of the same power with different numberings and so with different well-orderings. The numberings, then, depend

[29] Dauben, *Georg Cantor*, p. 101. The examples Dauben presents here and below come from the French version of Cantor's "Foundations of a General Theory of Manifolds;" Georg Cantor, "Fondaments d'une Théorie Générale des Ensembles," *Acta Mathematica* 2 (1883), p. 393.

[30] Ibid., p. 102.

upon the ordering of the elements so, in general, different orderings produce different numberings, although they have the same number of elements. In addition, there is a correlation between the number of elements in a set and the number of numberings the set could yield depending upon its ordering. For example, Cantor considered the sets of the first number class (I) given in a determinate order. The numberings of these sets, so long as they are well-ordered, always correspond to numbers of the second number class (II). Also, conversely, given any number α of (II), any set of (I) can be ordered so that its numbering would correspond to α. Analogous results hold, as well, for sets of higher power.[31]

As was indicated previously, when focused only on finite sets, the concepts of power and numbering coincide; different orderings of finite sets do not produce different powers. Since for finite sets power was independent of ordering, a finite set of n elements just had power n. But for infinite sets the distinction between power and number was significant. Every number α of (II) indicated a unique ordering of elements, but any such set with α as its numbering was always denumerable. Here, a connection remained between power and numbering in that any denumerable well-ordered set had a uniquely determined numbering α, where α belonged to the second number class (II). The properties of well-ordered sets had the result of joining these various notions of transfinite numbers, numberings, and number classes together into a unified theory. The concept of numbering merely generalized the notion of counting. For Cantor, this further supported his view that the transfinite numbers were just as legitimate as their finite counterparts.[32]

Kanamori explains that Cantor's interest in the well-ordering principle and the ordinal numbers, that is, the numberings, was directly linked to his attempts to solve the continuum problem.[33] The transfinite ordinals provided Cantor with a framework for his two primary approaches to the problem: one through power and the other through definable sets of real numbers. His approach through power focused on the first and second number classes – the natural numbers and the set of countably infinite ordinals respectively – although he also gave some indication that these classes could be continued to a third number class and beyond. His major result here was to prove that the number class (II) was uncountable and that any subset of (II) is either countable or of the same power

[31] Ibid., pp. 102–3.

[32] Ibid., p. 103.

[33] Kanamori, "The Mathematical Development of Set Theory from Cantor to Cohen," p. 6.

as (II). Thus the second number class (II) had precisely the property Cantor sought for the real numbers, reducing the continuum hypothesis to the assertion that the set of real numbers and the second number class were of the same power. However, he could not seem to find a definable well-ordering of the real numbers and so could not find a correlation between these two sets. Cantor had not yet developed his famous diagonal method of proof for showing (II) uncountable. Instead, in section 12 of the *Grundlagen*, he presented an argument similar to that given earlier in his proof that the real numbers are uncountable.

1.3 Cardinals, Ordinals, and the Continuum Problem

In the years following publication of the *Grundlagen*, Cantor continued to be unable to solve the continuum problem by searching for direct correlations between the set of real numbers and the ordinals. This led him to a more general approach to size and order that would take into account the continuum. To this end, he introduced the notion of a cardinal number to replace the earlier terminology of "power." He also went beyond the study of well-orderings to the more general notion of linear order types, and he took on a view of well-defined sets as being given together with a linear ordering of their members. Order types and cardinal numbers resulted from the successive abstraction from a set M to its order type $\overline{M}$ and then to its cardinal $\overline{\overline{M}}$. For ordinals the process is similar, but we do not also abstract away the ordering of the set.[34] It was in this context that Cantor put forward in his 1891 "On an Elementary Question in the Theory of Manifolds" his famous diagonal argument establishing that for any set M, the collection of functions from M into a two-element set is of higher power, or cardinality, than M itself.[35]

As we saw, Cantor proved already in 1874 the existence of uncountable sets. This proof, however, relied on the existence of irrational numbers. In 1891, he reproved this claim with much greater generality, appealing only to the abstract sets themselves. But in doing so, he also employed extremely powerful methods, leading him to an ascending and limitless hierarchy of transfinite powers. With this, he brought the

[34] Georg Candor, *Contributions to the Founding of the Theory of Transfinite Numbers* (1895, 1897), trans. Philip E.B. Jourdain (New York, NY: Dover Publications, Inc., 1955), pp. 86–112.

[35] Kanamori, "The Mathematical Development of Set Theory from Cantor to Cohen," p. 7; Cantor, "On an Elementary Question in the Theory of Manifolds," in *From Kant to Hilbert*, Ewald, pp. 920–2.

mathematical infinite to full maturity. Cantor's diagonal method allowed him to avoid all talk of point sets, irrational numbers, or any specific objects at all. In this way, he achieved complete generality, relying on collections of no specific sorts of potentially questionable entities at all, except perhaps his sets themselves.

The diagonalization proof relied on two elements m and w. Cantor used these to consider a collection M made up of elements $E = (x_1, x_2, \ldots, x_n, \ldots)$, where each x_n was either m or w. So, for example, the elements E might look like any of the following:

$$E^{\mathrm{I}} = (m, m, m, m, \ldots),$$
$$E^{\mathrm{II}} = (w, w, w, w, \ldots),$$
$$E^{\mathrm{III}} = (m, w, m, w, \ldots).$$

He then claimed that any such collection M was uncountable. If $E_1, E_2, \ldots, E_\nu, \ldots$ is any simply infinite sequence of elements of the set M, then there is an element E_0 in M which is not equal to any E_ν. To prove his claim, he first gave a countable listing of elements E_μ, where each $a_{\mu,\nu}$ was either m or w:

$$E_1 = (a_{11}, a_{12}, \ldots, a_{1\nu}, \ldots),$$
$$E_2 = (a_{21}, a_{22}, \ldots, a_{2\nu}, \ldots),$$
$$\ldots$$
$$E_\mu = (a_{\mu 1}, a_{\mu 2}, \ldots, a_{\mu\nu}, \ldots)$$
$$\ldots$$

This defined a new sequence $b_1, b_2, \ldots, b_\nu, \ldots$, where each b_ν was either m or w but $b_\nu \neq a_{\nu\nu}$. This sequence of b_ν yielded a new element $E_0 = (b_1, b_2, \ldots, \mathrm{b}_\nu, \ldots)$ in M, where $E_0 \neq E_\nu$ for any index ν. No matter what element E_ν was considered, E_0 always differed from it at the νth coordinate. Therefore there was always an element left off of a countable listing of elements of M, and hence M was uncountable.[36]

Cantor then proceeded to show how to apply his diagonal method to specific infinite sets, considering the set L of the linear continuum and the set of real numbers on (0, 1) and then showing that the set M of single-valued functions $f(x)$ with only values 0 or 1 for any x in (0, 1) was of greater power than L. M was clearly greater than or equal to L in

[36] Dauben, *Georg Cantor*, p. 166; Cantor, "On an Elementary Question in the Theory of Manifolds," pp. 920–1.

power, since it contained a subset equal in power to L. For example, the set N of functions $f(x)$ on (0, 1) equal to 0 everywhere except at a single point x_0, where $f(x_0) = 1$ was such a subset. So Cantor had to show only that L and M are in fact not equal in power. If M and L are equal in power, then it must be possible to establish a one-to-one mapping between them. That is, there must be a function $\varphi(x, z)$, where for every value z, there must be an element $f(x)$ in M such that $f(x) = \varphi(x, z)$. Conversely, for every element $f(x)$ in M given by $\varphi(x, z)$, there must be a unique z such that $f(x) = \varphi(x, z)$. This is impossible, as is shown by the diagonalization method. Consider the function $g(x)$ having only values 0 or 1 but where $g(x) \neq \varphi(x, x)$ for any given x. Hence $g(x)$ is in M, but z could not be determined in a way that would yield $g(x)$ from $\varphi(x, z)$, since $\varphi(z_0, z_0)$ is never equal to $g(z_0)$. Therefore M is of greater power than L.[37]

From his new proof of the existence of uncountable sets, Cantor provided a simple way for showing that the ascending sequence of powers of well-defined sets had no maximum. Indeed, he took this to be the true significance of the proof:

> This proof is remarkable not only because of its great simplicity, but more importantly because the principle therein can be extended immediately to the general theorem that the powers of well-defined manifolds have no maximum, or what is the same thing, that for any given manifold L we can produce a manifold M whose power is greater than that of L.[38]

Furthermore, Cantor took this infinite ascending sequence of higher and higher powers as a generalization of the notion of a finite cardinal and therefore as having a status equal to that of the finite numbers. As he remarked,

> The "powers" represent the unique and necessary generalization of the finite "cardinal numbers." They are none other than the actual-infinite cardinal numbers, and they have the same reality and determinateness as the others, except that the law-like relations among them – their "number theory" – is in part of a different sort than in the domain of the finite.[39]

For Cantor – looking at the history of mathematics – the rational, irrational, and complex numbers had been accepted because of their utility and consistency. Each of these number systems was a consistent

[37] Dauben, *Georg Cantor*, p. 167; Cantor, "On an Elementary Question in the Theory of Manifolds," p. 922.

[38] Ibid., pp. 921–2; Dauben also quotes this passage on p. 166 of *Georg Cantor*.

[39] Cantor, "On an Elementary Question in the Theory of Manifolds," p. 922.

generalization of prior less comprehensive concepts. The concept of power, then, was the most comprehensive and natural of all.[40]

Cantor's final major work in set theory was his 1895/1897 *Contributions to the Founding of the Theory of Transfinite Numbers*, commonly known as the *Beiträge*.[41] This work summarizes his progress in set theory to this point while also clearly revealing a gap left open by his inability to settle the continuum problem. In part I, he presents his post-*Grundlagen* research on cardinal numbers and the continuum, posing the question of cardinal comparability, whether for all cardinals a and b, either $a = b$, $a < b$, or $b < a$. He promised a proof at some point later but never provided one, and it did not come until it appeared as a consequence of Zermelo's 1904 well-ordering theorem. Putting comparability aside for now, however, Cantor first turned to defining the addition, multiplication, and exponentiation of cardinal numbers, showing that, like the finite numbers, the infinite too were governed by arithmetical laws. He also introduced the now standard aleph notation for infinite cardinals, taking $\aleph_0$ as the cardinality of the set of natural numbers, the first infinite cardinal number. He further observed that $2^{\aleph_0}$ is the cardinality of the set of real numbers and easily established, by way of his cardinal arithmetic, the one-to-one correspondence between $\mathbb{R}$ and $\mathbb{R}^n$, since $(2^{\aleph_0})^{\aleph_0} = 2^{\aleph_0 \cdot \aleph_0} = 2^{\aleph_0}$. Finally, he presented his theory of order types, taking η as the order type of the rational numbers, a countable dense linear order without endpoints, and θ as the order type of the real numbers, a perfect linear order with a countable dense set.[42]

In part II of *the Beiträge*, Cantor further developed his views on well-orderings construing their order types as the ordinals and proving, by way of order comparisons of well-ordered sets, that they are comparable. He then presented ordinal arithmetic as just a special case of the arithmetic of order types. He also paid some attention to the properties of the second number class (II) and defined its cardinal number as $\aleph_1$. In the final sections, he turned to ordinal exponentiation in the second number class, defining this operation by transfinite recursion and using it to establish his famous normal form theorem. He concluded with a discussion of the numbers satisfying the condition $\varepsilon = \omega^{\varepsilon}$, now known as the epsilon numbers.[43]

[40] Dauben, *Georg Cantor*, p. 167.

[41] Georg Cantor, *Contributions to the Founding of the Theory of Transfinite Numbers*, trans. Philip E. B. Jourdain (New York, NY: Dover Publications, 1955).

[42] Kanamori, "The Mathematical Development of Set Theory from Cantor to Cohen," p. 8; Cantor, *Contributions*, pp. 90, 91–7, 123, 133.

[43] Ibid., p. 9; Cantor, *Contributions*, pp. 173, 183–201.

With the first part of the *Beiträge* dealing primarily with cardinal numbers and the continuum and the second part focusing on ordinal numbers and well-orderings, the two parts remained distinct and lacking in unity in their subject matter. Kanamori states that this was not just an oversight but the indication of a serious split between the areas of Cantor's developing set theory.[44] For example, nowhere in part I does Cantor state, even as a special case, his major 1891 result that $a < 2^a$ for any set a, either finite or infinite. Instead, the succession of transfinite cardinals is taken as the alephs defined as the powers of the sets of their predecessor ordinals. But then in part II, Cantor never mentions any aleph beyond $\aleph_1$, or that the continuum hypothesis can be stated as $2^{\aleph_0} = \aleph_1$. Furthermore, he establishes ordinal comparability but is unable to reduce cardinal comparability to it. Kanamori concludes, "Having ushered in arbitrary functions through cardinal exponentiation Cantor had introduced an irreconcilable tension into his view that all sets are well-ordered, and there was little point to developing the theory of higher alephs with the assurance of their gauging all the cardinal numbers."[45] However, the 1904 International Congress of Mathematicians at Heidelberg signaled a turning point. Julius König presented a proof of the claim that $2^{\aleph_0}$ was not an aleph, meaning that the continuum was in fact not well-orderable. Although by the next day, Zermelo had shown König's proof to be faulty, Cantor was shaken with worries that the continuum might escape the context for quantifying it that he had worked to construct for this very purpose.[46] This was not the only problem looming for Cantor's set theory. By this time, the paradoxes of set theory were also threatening, and it is to this topic that I turn next.

[44] Ibid., p. 8.
[45] Ibid., pp. 8–9.
[46] Ibid., p. 9.

CHAPTER 2

Cantor, Russell, and Zermelo and the Set-Theoretic Paradoxes

In this chapter, I turn to consider the set-theoretic paradoxes in their historical and philosophical context. There has been much recent scholarship in this area, attempting to sort out the rather complicated history of how the paradoxes arose, when they were discovered, and by whom. Earlier accounts of this history had Cesare Burali-Forti discovering the first of the paradoxes – that of the largest ordinal, which became the Burali-Forti paradox – with the Cantor and Russell paradoxes following soon after.[1] More recently, this story has seen considerable revision, particularly at the hands of Gregory Moore and Alejandro Garciadiego.[2] Specifically, they have argued that the discovery of the earliest set-theoretic paradoxes – Burali-Forti's, Georg Cantor's, and Bertrand Russell's – can all be traced to Russell himself.[3] Following them, I pick up the story of the paradoxes with Russell.

In tracing the emergence of the Burali-Forti paradox to Russell, Moore and Garciadiego point to Russell's early Hegelianism as its source, explaining, "Russell began to search for paradoxes in mathematics much earlier than is usually recognized. His predisposition to invent such paradoxes had its roots in the philosophical antinomies of Kant and Hegel,

[1] For example, see Jean van Heijenoort's introductory note to Cesare Burali-Forti, "A Question on Transfinite Numbers," in *From Frege to Gödel: A Source Book in Mathematical Logic, 1879–1931*, ed. Jean van Heijenoort (Cambridge, MA: Harvard University Press, 1967), pp. 104–11.

[2] Gregory H. Moore and Alejandro Garciadiego, "Burali-Forti's Paradox: A Reappraisal of Its Origins," *Historia Mathematica* 8 (1981), pp. 319–50; Alejandro Garciadiego, *Bertrand Russell and the Origins of the Set-Theoretic "Paradoxes"* (Boston, MA: Birkhäuser Verlag, 1992). They note that van Heijenoort was far from the only historian to hold this view, citing, among others, the similar attitudes of I. M. Bochenski, Nicolas Bourbaki, Morris Kline, and William and Martha Kneale; see p. 320 of Moore and Garciadiego, "Burali-Forti's Paradox," and pp. 19–20 of Garciadiego, *Bertrand Russell and the Origins of the Set-Theoretic "Paradoxes."*

[3] The last of these needs some qualification. Ernst Zermelo had discovered Russell's paradox independently – and before Russell himself – but did not publish it. For more on Zermelo's discovery, see B. Rang and W. Thomas, "Zermelo's Discovery of the 'Russell Paradox.'" *Historia Mathematica* 8 (1981), pp. 15–22.

both of whom deeply influenced his early development as a philosopher."[4] Adding to this story, I would emphasize that Russell came upon the set-theoretic paradoxes, including the one that would bear his name, by considering some fairly intuitive but unfortunately contradictory features that we might expect the notion of a set to have. This aspect in his discovery of the paradoxes will be especially important throughout much of what follows in this and later chapters. It brings out that the foundational crisis in mathematics was a battle of intuitions about sets and that its resolution ultimately came down to largely pragmatic considerations which, I will claim, reveals a distinct tradition of explication in the philosophy of set theory. The paradoxes showed that a single notion of set, incorporating all of the intuitive features that both mathematical and philosophical logicians initially thought sets to have, was impossible. As a result, set theory's founders looked to the mathematically significant features of set theory that were traditionally conceived to give an explicative account of the notion of set rather than a uniquely correct conceptual analysis of the entities that set theory was supposed to be about.

2.1 Russell and the Discovery of the Paradoxes

Russell had arrived at his paradox by May 1901, which he discovered, according to his account in *Principles of Mathematics*, "in the endeavor to reconcile Cantor's proof that there can be no greatest cardinal number with the very plausible supposition that the class of all terms … has necessarily the greatest possible number of members," a term being "[w]hatever may be an object of thought, or may occur in any true or false proposition, or can be counted as *one*."[5] Moore puts forth this date for the discovery based upon an early manuscript of *Principles* where in chapter three, Russell composes the first extant account of his paradox:

> The axiom that all referents with respect to a given relation form a class seems, however, to require some limitation, and that for the following reason. We saw that some predicates can be predicated of themselves. Consider now those (and they are the vast majority) of which this is not the case. These are the referents (and also the relata) in a certain complex relation, namely the combination of non-predicability with identity.

[4] Moore and Garciadiego, "Burali-Forti's Paradox," p. 324.

[5] Bertrand Russell, *Principles of Mathematics*, 2nd edn (London: George Allen and Unwin, 1937), p. 101. The first edition of *Principles of Mathematics* appeared in 1903. Russell defines the terms on p. 43. He makes the same claim in his later reminisces; see Bertrand Russell, *My Philosophical Development* (New York, NY: Routledge, 1993 [1958]), p. 58.

> But there is no predicate which attaches to all of them and to no other terms. For this predicate will either be predicable or not predicable of itself. If it is predicable of itself, it is one of those referents by relation to which it was defined, and therefore, in virtue of their definition, it is not predicable of itself. Conversely, if it is not predicable of itself, then again it is one of the said referents, of all of which (by hypothesis), it is predicable, and therefore again it is predicable of itself. This is a contradiction, which shows that all the referents considered have no common predicate and therefore do not form a class ... It follows from the above that not every definable collection of terms forms a class defined by a common predicate. This fact must be borne in mind, and we must endeavor to discover what properties a collection must have in order to form a class.[6]

Here, we see not only Russell's first statement of the paradox, but he also gives the first indication of where its resolution will ultimately lie. In observing that "the axiom that all referents with respect to a given relation form a class" needs limiting, Russell suggests the traditional conception of what a class is: that every predicate or concept determines a class. As we have seen and will continue to see, this idea of a set was found throughout the emerging discipline of set theory, including that work of Cantor. Locating the source of the paradox in the comprehension principle changed the development of set theory in crucial ways and, in particular, gave rise to the philosophical debates over set theory that we will primarily be concerned with in many of the following chapters. I should also note that in moving toward rejecting full comprehension, Russell does not completely reject the idea that there is a uniquely correct version of set in favor of a more pragmatic conception, that is, Russell does not completely move from the conceptual analysis view to the explication view. (I do not think Russell ever adopts one view or the other completely, which often leads to tensions in his philosophy of mathematics.) Instead, as Russell remarks at the end of the passage, now the task becomes "to discover what properties a collection must have in order to form a class." He has not yet suggested that the mathematical demands of set theory can settle this question.

Russell did not, however, react with immediate concern. The available textual evidence suggests that he reported the paradox to no one else – except Giuseppe Peano, who never responded – until June 1902, when

[6] Bertrand Russell, *The Collected Papers of Bertrand Russell*, vol. 3 (New York, NY: Routledge, 1990), p. 195. A version of this passage is quoted in Gregory H. Moore, "The Origins of Russell's Paradox: Russell, Couturat, and the Antinomy of Infinite Number," in *Essays on the Development of the Foundations of Mathematics*, ed. Jaakko Hintikka (Dordrecht: Kluwer Academic Publishers, 1995), pp. 233–4.

he communicated it to Gottlob Frege. At that point, Frege took it as a matter of great concern:

> Your discovery of the contradiction caused me the greatest surprise and, I would almost say, consternation, since it has shaken the basis on which I intended to build arithmetic. It seems, then, that transforming the generalization of an equality into an equality of courses-of-values (§ 9 of my *Grundgesetze*) is not always permitted, that my Rule V (§ 20, p. 36) is false, and that my explanations in § 31 are not sufficient to ensure that my combinations of signs have a meaning in all cases. I must reflect further on the matter. It is all the more serious since, with the loss of my Rule V, not only the foundation of my arithmetic, but also the sole possible foundations of arithmetic, seem to vanish.[7]

Moore reasonably suggests that Russell was not initially disturbed by his paradox because he thought there must be some simple solution to it. Only upon receiving Frege's pessimistic response did Russell see how devastating the consequences of his discovery were. He looked upon Frege with extremely high regard, and when he could offer no simple resolution, Russell could no longer see this as a minor difficulty that would be solved in time.[8] For Russell, the foundational crisis had arrived.

From this point, matters became worse. Moore and Garciadiego point to Russell's recognition of the seriousness of his own paradox as the source of the other set-theoretic paradoxes. Now, reconsidering the work of Burali-Forti and of Cantor, Russell came to recognize also the two paradoxes that would eventually bear their respective names. In a 1902 article on well-ordering, Russell began his attempts at reconciling the works of Burali-Forti and Cantor, now accepting Cantor's proof of the comparability of ordinals but doubting his claim that every set can be well ordered. Russell's response at this point was to grant that every ordinal segment was well-orderable but not that the ordinals as a whole were. In this way, he continued to avoid the conclusion that Burali-Forti and Cantor had landed themselves in contradiction.[9]

This situation changed in the 1903 publication of *Principles of Mathematics*. Here, in addition to an entire chapter on his own paradox, Russell put forward, for the first time in print, both the Burali-Forti and

[7] Gottlob Frege, "Letter to Russell" (1902), in *From Frege to Gödel*, ed. van Heijenoort, pp. 127–8. I have followed the translation given there. The "Rule V" that Frege refers to is what is more commonly known as his "Basic Law V."

[8] G. H. Moore, "The Origins of Russell's Paradox," pp. 234–5; see also Moore and Garciadiego, "Burali-Forti's Paradox," pp. 328–9.

[9] Moore and Garciadiego, "Burali-Forti's Paradox," pp. 329–30.

Cantor paradoxes.[10] As already mentioned previously, Russell claims to have discovered his own paradox in considering Cantor's theorem. In chapter ten of *Principles of Mathematics*, entitled "The Contradiction," he gives extensive treatment to the paradoxes, rehearsing his own in three different forms. In diagnosing the problem, he again turns to the comprehension principle, the axiom that any propositional function in one free variable determines a class. He observes that either this principle or the principle that every class can be treated as a single term must be false, but he sees no fundamental objection to rejecting either. "But having dropped the former," Russell asks, "[w]hich propositional functions define classes which are single terms as well as many, and which do not? *And with this question our real difficulties begin* [my emphasis]."[11] This, as we will continue to see, is indeed the crucial issue to set theory in the early twentieth century and, in a sense, remains to this day one of the crucial issues for set theory in its more philosophical aspects. Having spelled out his own paradox, Russell now returns to Burali-Forti and Cantor, seeing for the first time that their views also lead him into difficulty, though of a less general nature than the one that emerges from his own paradox.

Russell first considers Burali-Forti, remarking that there appears to be a problem with the order type of the whole series of ordinals. We can easily prove that every segment of this series is well ordered, so it is natural to suppose that the whole series is also well ordered. The order type of this series would be the greatest of all the ordinals, since the ordinals that are less than a given ordinal form, in order of increasing magnitude, a series whose order type is the given ordinal. But every ordinal can be increased by the addition of 1, so there can be no greatest ordinal. On the basis of this contradiction, Russell correctly observes that Burali-Forti concluded that for any two given ordinal numbers, comparability does not in general hold. Russell also observes that Cantor proved exactly the opposite.[12] Here, we have Russell's discovery of the Burali-Forti paradox.

Russell finds no weakness in Cantor's contrary proof. Instead, he questions Burali-Forti's premise that the whole series of ordinals is well ordered. From the well-ordering of every segment of ordinals, Russell

[10] Some scholars have traced the Cantor paradox to the 1899 Cantor–Dedekind correspondence, but as we will see, this is not entirely accurate. Cantor does produce a contradiction there but, much like Burali-Forti, only in the service of a reductio proof. It was Russell who turned these contradictions into a more general foundational crisis for mathematics.

[11] Russell, *Principles of Mathematics*, pp. 102–3.

[12] Ibid., p. 323.

claims that the well-ordering of the whole series does not follow. Because this claim of the well-ordering of the entire series of ordinals appears incapable of proof and rejecting it avoids Burali-Forti's paradox, Russell follows this path and, at least for now, takes this to be the successful resolution of the paradox of the largest ordinal.[13] Russell does not see, however, that Cantor himself nowhere explicitly admits a class of all ordinals, and that without this, there is no contradiction. What drives Russell to the paradox is his perhaps not unreasonable view that the ordinals themselves should form a class. Had there been no paradox, it seems no more surprising to have a class of all ordinals than to have a class of all natural or real numbers.

In introducing his final set-theoretic paradox, Russell turns to Cantor himself, considering his investigations into the cardinal numbers. Russell begins by remarking that the common objections to infinite numbers, classes, series, and the infinite in general as self-contradictory are groundless, but more serious problems remain, related to his own paradox of non-self-membered classes. This is not a problem with the infinite as such, he argues, but only with certain large infinite classes, explaining the difficulty as follows. Cantor proved that there is no greatest cardinal number: If *u* is a class, then the number of classes included in *u* is greater than the number of terms, or elements, in *u*. In other words, for any cardinal α, $2^\alpha > \alpha$, which is Cantor's theorem. But Russell observes also that certain classes appear to have as many terms as possible, such as the class of all terms, the class of all classes, and the class of all propositions. So it appears to Russell that Cantor's proof must make some assumption that does not hold for such classes, since definite contradictions arise in applying Cantor's reasoning to these large classes. Indeed, such difficulties arise generally for any class of all entities or any equally numerous class. Russell admits that in light of this problem, we might be tempted to think that the totality of things, or the whole universe of existing entities, is an illegitimate totality, inherently contradictory to logic. "But it is undesirable to adopt so desperate a measure," he counters, "so long as hope remains of some less heroic solution."[14]

Russell concludes these investigations by observing that although Cantor's argument leads to contradictions where such large classes are concerned, he can find no missteps in Cantor's original proof. The only solution he sees at this point is to deny that there are any true

[13] Ibid.
[14] Ibid., p. 362.

propositions concerning all objects or propositions. "Yet the latter, at least, seems plainly false," he says, "since all propositions are at any rate true or false, even if they have no other common properties. In this unsatisfactory state, I reluctantly leave the problem to the ingenuity of the reader."[15] For Cantor, by contrast, we will see that it seems plainly true that there is no universal class. I will now consider how these difficulties were in fact dealt with by Russell and others.

2.2 Avoiding the Paradoxes: Cantor and the Absolute Infinite

I turn first to Cantor, both as the founder of set theory and as someone who may never have been confronted by problems such as those that Russell introduced. For while Russell saw it as near common sense that such large classes as the class of all classes and the class of all objects should exist, Cantor saw this as equally implausible. As we will see, in Cantor's view of the nature classes, we are unable to derive such paradoxes. The only contradictions Cantor derives are the sorts one would ordinarily and unsurprisingly expect to find in reductio proofs.[16] The first occurrence of Cantor's paradox has often been traced to his 1899 correspondence with Richard Dedekind, but this interpretation presents two problems: the one just mentioned and the other that Cantor's paradox was not publicly known before 1932 when the relevant pieces of the Cantor–Dedekind correspondence were published. There are, however, still interesting and relevant aspects of this correspondence for the set-theoretic paradoxes, in particular for understanding how they were resolved.

Cantor initiated his 1899 correspondence with Dedekind at the end of July, writing that he would like them to correspond regularly so that he might have Dedekind's views on certain fundamental questions of set theory.[17] In particular, Cantor wrote to Dedekind on August 3, presenting his proof that every transfinite set has a definite cardinal and identifying the system ת (tav) of all alephs with the system of all transfinite cardinal numbers, that is, all transfinite cardinals are alephs. But before proving this result, Cantor introduces what he sees as a crucial

[15] Ibid., p. 368.

[16] This is as close as Burali-Forti ever came to paradox as well, as is shown in his "A Question on Transfinite Numbers," pp. 105–11. Here, he does not show that Cantor's theory is inconsistent but, rather, attempts to prove the trichotomy law for ordinals by way of a reductio proof. There is a contradiction in the paper but only the one needed for the reductio.

[17] Georg Cantor, Richard Dedekind, and David Hilbert, "Cantor's Late Correspondence with Dedekind and Hilbert," in *From Kant to Hilbert,: A Source Book in the Foundations of Mathematics*, vol. II, ed. William Ewald (Oxford: Clarendon Press, 1999), pp. 923–40.

distinction: that between consistent and inconsistent multiplicities. He explains that he begins with the notion of a definite multiplicity (or system or totality) of things but then observes that the assumption that a multiplicity's elements "are together" can lead to a contradiction in certain cases. So contrary to our original thought, such a multiplicity cannot be conceived of as a unity, or "one finished thing," and these multiplicities, he says, are the absolutely infinite, or inconsistent, multiplicities. He observes that "[a]s we can readily see, the 'totality of everything thinkable', for example, is such a multiplicity; later still other examples will turn up."[18] Of course, as we saw, Russell did not readily see this, and his initial commitment to such classes is what brought him to the paradoxes. The idea that I will begin to develop here and advance more in future chapters is that the systemization of set theory has largely been, and remains, a battle of competing intuitions about what sets are like. As we will see, what set theory's founders needed was a principled way for determining which sets exist that would also eliminate the paradoxes.

In contrast to his inconsistent multiplicities, Cantor also observes that there are collections in which the elements can be thought as "being together" without contradiction and so can be gathered into "one thing." These are the consistent multiplicities, or sets. He goes on to note that if two multiplicities are equivalent, then they are both either sets or inconsistent and every submultiplicity of a set is also a set. Also, for any set of sets, the elements of these sets also form a set.[19] Using this distinction between consistent and inconsistent multiplicities, Cantor begins his proof that every transfinite cardinal is an aleph. First, he considers the system Ω of all ordinals, noting that he had earlier proved that any two ordinals are comparable.[20] This multiplicity, when naturally ordered according to magnitude, forms a well-ordered sequence

$$0, 1, 2, 3, \ldots, \omega_0, \omega_0 + 1, \ldots, \gamma, \ldots$$

in which every number is the order type of the preceding sequence of elements. Next, he shows that Ω is not a consistent multiplicity. Assume that Ω is consistent. Then, since it is well ordered, it corresponds to some ordinal δ greater than all the ordinals in Ω. But δ is also in Ω, since Ω contains all the ordinals. So we have $\delta < \delta$, a contradiction. Therefore Ω is an inconsistent, or absolutely infinite, multiplicity.

[18] Ibid., pp. 931–2.

[19] Ibid., p. 932.

[20] Cantor actually considers Ω', which he takes as the ordinals plus zero. For simplicity, I will just take the ordinals to include zero.

Cantor then further explains that since the similarity of well-ordered sets also yields their equivalence – that is, they have the same cardinality – to every ordinal γ there corresponds a definite cardinal $c = \overline{\gamma}$ where $\overline{\gamma}$ is the general notion applying to all ordered sets similar to γ. That is, c is the cardinal of any well-ordered set with type $\overline{\gamma}$. As we saw, Cantor denotes the cardinals corresponding to the transfinite ordinals by "$\aleph_n$," aleph, the first letter of the Hebrew alphabet. He explains to Dedekind that the system of ordinals γ corresponding to one and the same cardinal **c** forms "a number class," Z(c). In every such number class, there is a least ordinal γ_0, and there is also a first ordinal γ_1 falling outside this number class Z(c) such that the condition $\gamma_0 < \gamma < \gamma_1$ is equivalent to the fact that γ belongs to Z(c). Therefore every number class is a definite segment of the sequence Ω. Finally, Cantor notes that certain ordinals of the system Ω each by itself forms a number class. For example, to the finite ordinals

$$0, 1, 2, 3, \ldots, \nu, \ldots,$$

there correspond the finite cardinals

$$\overline{0}, \overline{1}, \overline{2}, \overline{3}, \ldots, \overline{\nu}, \ldots$$

Now let ω_0 be the least transfinite ordinal, and let $\aleph_0$ be the cardinal corresponding to it, so that $\aleph_0 = \overline{\omega}_0$. So $\aleph_0$ is the least aleph and determines the number class $Z(\aleph_0) = \Omega_0$. The ordinals α of Ω_0 satisfy the condition $\omega_0 < \alpha < \omega_1$ and are characterized by it, and ω_1, then, is the least transfinite ordinal whose cardinal does not equal $\aleph_0$. So $\omega_1 = \aleph_1$, and $\aleph_0 \neq \aleph_1$, the next greater aleph, and so on. Cantor concludes by observing that among the transfinite ordinals of Ω, there is a least ordinal to which no $\aleph_\nu$ corresponds, where ν is finite, which he denotes as "ω_{ω_0}" This yields an aleph $\aleph_{\omega_0}$ which is the next greatest cardinal after all the $\aleph_\nu$. This process of forming alephs and number classes of Ω "that correspond to them is *absolutely* limitless" [Cantor's italics].[21] The system of all alephs he denotes by "л," tav, the twenty-second letter of the Hebrew alphabet, and he can now show that it too is an inconsistent, or absolutely, infinite totality. He observes that the system л of all alephs that are ordered according to magnitude,

$$\aleph_0, \aleph_1, \ldots, \aleph_{\omega_0}, \aleph_{\omega_0+1}, \ldots, \aleph_{\omega_1}, \ldots,$$

forms a sequence similar to the system Ω and so is also inconsistent.[22]

[21] Cantor, Dedekind, and Hilbert, "Cantor's Late Correspondence," pp. 933–4.

[22] Ibid., p. 934.

Cantor's final result in this letter answers the question "Are there transfinite cardinals not contained in the system ℶ?" He answers negatively on the basis of the inconsistency of Ω and ℶ. If we define a multiplicity *V* and assume that no cardinal corresponds to it, then *V* must be inconsistent. By this assumption, Ω must be projectible into *V*, so there must be a submultiplicity *V′* of *V* equivalent to Ω. *V′*, then, is inconsistent, since Ω is, so the same must hold of *V*. Therefore, he concludes, every consistent transfinite multiplicity must have a definite aleph as its cardinal, so all transfinite cardinals are contained in the system ℶ. From this, he also concludes that his 1895 assertion of the comparability of cardinals was in fact correct.[23]

In an editorial note to this letter, Ernst Zermelo observes a weakness in Cantor's proof, namely, that Cantor had not proved that Ω was in fact projectible into every *V* lacking an aleph as its cardinal. Something like an intuition of time seemed to be at play here, with Cantor continuing a process of successively and arbitrarily matching elements of *V* with the ordinals of Ω. However, this assumes that each of the elements of *V* will be used only once in the process, and such a process, Zermelo remarks, "goes beyond all intuition." What Cantor needed was Zermelo's axiom of choice so that he could make a simultaneous choice of elements from *V* to define *V′*.[24] It was this axiom that eventually allowed Zermelo to prove what Cantor had wanted.[25]

Although we will return to the axiom of choice in later chapters, this weakness in Cantor's proof does not obscure my main concern here. What is important to note is that while something like Cantor's paradox does appear in this letter to Dedekind, it is all part of a reductio proof.[26] Cantor's paradox did not result in a foundational crisis because this contradiction was what Cantor intended to prove for reductio so that he could obtain his results about the transfinite cardinals. Why, then, did Russell react so differently to the paradoxes? This is the second important point in Cantor's letter. In his distinction between consistent and inconsistent totalities, Cantor had a response to the paradoxes before any foundational crisis arose. But what were the grounds for such a distinction? It is not exactly clear when Cantor introduced it, although the earliest

[23] Ibid., pp. 934–5.

[24] Ibid., p. 935, fn. b.

[25] Ernst Zermelo, "Proof That Every Set Can Be Well-Ordered" (1904) in *From Frege to Gödel*, ed. van Heijenoort, pp. 139–41. Owing to controversy over his use of the axiom of choice, Zermelo reproved the result, again using AC, in 1908; see "A New Proof of the Possibility of a Well-Ordering," in *From Frege to Gödel*, ed. van Heijenoort, pp. 183–98.

[26] Moore and Garciadiego emphasize this point in their work.

evidence for some version of it is an 1897 letter to David Hilbert.[27] One possibility is that Cantor did in fact first discover the set-theoretic paradoxes and then introduced it as a solution to them. If this is the case, Cantor's solution to the paradoxes was very much on a par with the solutions offered later by Russell and Zermelo. As we will see, they were all in a sense introduced on pragmatic grounds with the aim of preserving as much of the mathematical content of Cantor's original theory as possible. There does not seem to be much evidence for this as a direct account of Cantor's thinking on the paradoxes. However, this way of thinking about resolving the paradoxes is consistent with the mathematical attitude toward developing set theory that we examined in Chapter 1. There, we saw Cantor developing his theory in response to his mathematical aims. In a sense, we will see both Zermelo and Russell adopting this attitude in the service of coming to terms with the paradoxes themselves.

Another possibility for grounding Cantor's appeal to consistent and inconsistent totalities – and one that would be much more welcome to set theorists who think there is a single correct, or intended, notion of a set – is that Cantor in fact thought there was something inherent in what a set is that called for such a distinction. There does seem to be some evidence for this view. But it will not be especially satisfying to most set theorists. Joseph Dauben has shown that Cantor worried greatly over whether his views of the infinite were consistent with Catholic dogma – whether knowledge of the infinite was to claim the kind of knowledge that only God could have. The acknowledgment of absolutely infinite totalities alleviated all such worries, since their inconsistency showed that we did not in fact have knowledge of such infinities. Human understanding was limited to infinities that fell short of Ω or ת or the system of all things thinkable.[28]

[27] There is some earlier evidence for the distinction in Cantor's discussion of the Absolute; see for example Cantor, *Foundations*, pp. 916–17, fn. 2. I believe, though, that the 1897 letter may be the earliest evidence of Cantor's connecting the distinction explicitly to the set-theoretic paradoxes.

[28] Joseph Warren Dauben, *Georg Cantor: His Mathematics and Philosophy of the Infinite* (Princeton, NJ: Princeton University Press, 1979), Chapters 10 and 11. Dauben provides various references for Cantor's philosophico-theological views of set theory, but for his published views, see especially "Mitteilungen zur Lehre vom Transfiniten" (1887, 1888). In *Gesammelte Abhandlungen: Mathematische und Philosophischen Inhalts*, ed. Ernst Zermelo (Hildesheim: Georg Olms Verlagsbuchhandlung, 1962), pp. 378–439. Hallett, in his *Cantorian Set Theory and Limitation of Size* (Oxford: Clarendon Press, 1984), also discusses the relationship between Cantor's philosophy of set theory and his theological views; see especially pp. 35–7 and 43–5.

While the possibly ad hoc nature of such a distinction between consistent and inconsistent totalities did not bother the likes of Hilbert and Dedekind, the unclarity of it did. As was noted previously, the earliest explicit mention of the distinction came from a letter of September 26, 1897, to Hilbert communicating an early version of Cantor's proof that all transfinite cardinals are alephs. Just as in his letter to Dedekind, Cantor then uses a reductio proof to show that all transfinite cardinals are alephs. While we do not have Hilbert's responses to Cantor, Cantor's letter of October 2, 1897, suggests that Hilbert did not fully grasp the distinction Cantor wished to make. Cantor begins this letter by immediately observing that Hilbert overlooked the characteristic of being a finished set. He goes on to again explain his proof, remarking:

> One must only understand the expression "finished" correctly. I say of a set that it can be thought of as *finished* (and call such a set, if it contains infinitely many elements, "transfinite" or "super-finite") if it is possible without contradiction (as can be done with finite sets) to think of *all its elements as existing together*, and so to think of the set itself as *a compounded thing for itself*; or (in other words) if it is *possible* to imagine the set as *actually existing* with the totality of its elements [Cantor's italics].

He then notes that "the 'transfinite' coincides with what has since antiquity been called the 'actual infinite.'"[29] He explains that what he intends by a set is "an 'assembling together' [which] is only possible if an '*existing together*' is *possible* [Cantor's italics]." He concludes by contrasting this with the absolutely infinite sets:

> Infinite sets such that the *totality* of their elements cannot be thought of as "existing together" or as a "thing for itself" ... and that therefore also *in this totality* are absolutely not an object of further *mathematical* contemplation, I call "*absolutely infinite* sets", and to them belongs the "set of all Alephs" [Cantor's italics].[30]

We do not know whether Hilbert ever understood Cantor's distinction, but we do know that Dedekind was equally puzzled by it. In his response to Cantor's proof that every transfinite cardinal was an aleph, Dedekind expressed his confusion:

> You will certainly sympathize with me if I frankly confess that, although I have read through your letter of 3 August many times, I am utterly

[29] Georg Cantor, Richard Dedekind, and David Hilbert, "Cantor's Late Correspondence with Dedekind and Hilbert," In Ewald, *From Kant to Hilbert*, pp. 923–40.

[30] Ibid., p. 928.

> unclear about your distinction into consistent and inconsistent; I do not know what you mean by the "coexistence of all elements of a multiplicity", and what you mean by its opposite. I do not doubt that with a more thorough study of your letter a light will go on for me; for I have great trust in your deep and perceptive research.[31]

Cantor was not afraid to bring certain mentalistic elements into his set theory, but the precise mathematical minds of Dedekind and Hilbert were unable to comprehend what exactly Cantor's informal distinction came to, that is, how it was to yield objects precise enough for mathematical contemplation.[32] There seem to be both physicalistic and mentalistic aspects of Cantor's view of a set. He speaks of collecting together elements into a set as a sort of pseudo-physical process that works much like the collecting together of any other group of physical objects. So long as we can imagine such a process, we have an existing together of elements and so a consistent totality, or set. However, a group of physical objects existing together is much easier to make sense of than, say, the cardinal or ordinal numbers. For physical objects, we might just stand them all together, and then we see that they can in fact exist simultaneously together. Having actually done just such a thing many times, we do not find it hard to imagine such a process when we literally cannot bring these objects all together to one place. Why, then, would something analogous fail for all of the cardinals? It is not immediately obvious why we cannot imagine these objects existing together, as Cantor says that we cannot. The set of all cardinal numbers turns out inconsistent in Cantor's view, so in a sense maybe we cannot think of these objects as existing together. But it is unclear precisely why the analogy with other sorts of objects breaks down. For Dedekind, Hilbert, and Russell, who were not already inclined toward Cantor's distinction, the analogy of collecting things together in the mind did not help. What was needed was some mathematically precise and principled way of determining which sets exist that would also banish the paradoxes.

Looking at Cantor's published views on the notion of a set yield no further information as to what principle decided between consistent and

[31] Ibid., p. 937.

[32] For a defense of Cantor's views, see William Tait, "Frege Versus Cantor and Dedekind: On the Concept of Number," in *The Provenance of Pure Reason: Essays in the Philosophy of Mathematics and Its History* (New York, NY: Oxford University Press, 2005), pp. 212–51.

inconsistent totalities. In his *Contributions to the Founding of the Theory of Transfinite Numbers* published in 1895, Cantor defines a set as "any collection into a whole M of definite and separate objects m of our intuition or our thought."[33] Thus he relies on an apparently mentalistic collecting together. Earlier, in a footnote to the *Grundlagen* of 1883, Cantor did elaborate some this idea of a collecting together:

> Theory of manifolds. I use this word to designate a very broad theoretical concept which I have hitherto used only in the special form of a theory of geometric or arithmetical sets. In general, by a "manifold" or "set" I understand every multiplicity which can be thought of as one, i.e. every aggregate of determinate elements which can be united into a whole by some law.[34]

Here again, Cantor appeals to an ability of our thought as determining which sets exist. But he does give us something further in this passage in stating that the sets are those aggregates whose "elements … can be united into a whole by some law." But this somewhat more precise account of what a set is does not make Cantor's distinction any clearer. As we have seen, the comprehension principle alone does not determine which collections are consistent and which are not. Indeed, it was precisely comprehension that led Russell to the contradictions. So again we are left only with Cantor's mentalistic intuitions about which sets exist. We have here just a dispute of intuitions over what sets are like, while what was needed was some definite criterion for set existence that did not lead to contradictions in the theory. This is just what Zermelo and Russell would provide.

2.3 Resolving the Paradoxes: Zermelo and Russell

In his 1908 "Investigations into the Foundation of Set Theory I," Zermelo presented an axiom system for set theory that was "to exclude all contradictions … [while] retain[ing] all that is valuable in this theory."[35] While he does not spend nearly as much time directly considering the paradoxes as Russell does, Zermelo does state the aim of his article as removing such difficulties from the theory. In opening the

[33] Georg Cantor, *Contributions to the Founding of the Theory of Transfinite Numbers (1895, 1897)*, trans. Philip E. B. Jourdain (New York, NY: Dover, 1955), p. 85.

[34] Cantor, *Foundations*, 916, note 1.

[35] Ernst Zermelo, "Investigations into the Foundations of Set Theory I," in *From Frege to Gödel*, ed. van Heijenoort, p. 200.

article, he first describes the important foundational work that set theory makes possible, describing the theory as

> that branch of mathematics whose task is to investigate mathematically the fundamental notions of "number", "order", and "function", taking them in their pristine, simple form, and to develop thereby the logical foundations of all arithmetic and analysis; thus it constitutes an indispensable component of the science of mathematics.[36]

He then immediately raises the issue of whether the theory can remain a viable path for foundational research in light of the paradoxes:

> At present, however, the very existence of this discipline seems to be threatened by certain contradictions, or "antinomies", that can be derived from its principles – principles necessarily governing our thinking, it seems – and to which no entirely satisfactory solution has yet been found.[37]

Gregory Moore has argued that despite these remarks on the paradoxes, Zermelo's primary concern in his axiomatization is to give a precise account of his proof of the well-ordering theorem. The proof had been the subject of much controversy within the mathematical community since its original appearance in 1904.[38] And in 1908, before publishing his axiomatization, Zermelo provided a second proof of the theorem, which includes responses to his various critics.[39] Moore describes "Investigations into the Foundations of Set Theory" as a companion paper aimed at making explicit exactly which principles the well-ordering proof relied upon.[40] Such emphasis has its place in correcting the historical record, since the more standard interpretation has overemphasized the philosophical roots of set theory and the paradoxes. But a danger arises in this revised interpretation as well. Too much emphasis on the mathematical approach to set theory and with it the view that mathematical set theorists were not greatly troubled by the paradoxes may lead to the mistaken view that there were always two notions of a set, the mathematical and the philosophical or logical, and that the mathematical

[36] Ibid., p. 200.

[37] Ibid.

[38] The controversy arose particularly with regard to Zermelo's use of the axiom of choice. I will return to this topic in a later chapter. The original proof appears in Zermelo's "Proof That Every Set Can Be Well-Ordered," pp. 139–41.

[39] Ernst Zermelo, "A New Proof of the Possibility of a Well-Ordering," in *From Frege to Gödel*, ed. van Heijenoort, pp. 183–98.

[40] Gregory H. Moore, "The Origins of Zermelo's Axiomatization of Set Theory," *Journal of Philosophical Logic* 7 (1978), pp. 307–29. Moore's *Zermelo's Axiom of Choice: Its Origins, Development, and Influence* (New York, NY: Springer, 1982) is a book-length treatment of these issues.

conception was never contradictory.[41] While I want to concede that mathematicians often reacted differently to the paradoxes than more philosophically minded logicians, I also want to stress that all researchers in set theory at the beginning of the twentieth century had to come to terms with the paradoxes and that it was only in doing so that two conceptions of set emerged. To the extent to which there was any agreed-upon notion of a set prior to the paradoxes, it was roughly that a set is the extension of a predicate or a concept. Both mathematical and philosophical logicians had to clarify this notion of a set in light of the paradoxes. This is precisely how Zermelo goes on to describe his axiomatization.

As we saw in the passage from Zermelo, he describes the existence of contradictions in set theory as producing only a seeming threat to the discipline, indicating that he does not think the paradoxes are an actual threat. We should not take this to show that Zermelo never thought there was a threat from the paradoxes. The threat is only apparent now because there have been solutions to the paradoxes. We have seen already Cantor's attempt to resolve the paradoxes through his distinction between consistent and inconsistent totalities. Zermelo himself had already in his original 1904 well-ordering proof appealed only to principles, which he believed yielded no contradictory results.[42] Now his 1908 axiomatization can be added to this list. Still, he describes the current situation in set theory as being that no completely satisfying solution to the paradoxes has been found. We should not take this as indicating that some better solution is yet to be found. Rather, Zermelo's remark here indicates his recognition of the limitations that the paradoxes have imposed on set theory.

From Russell's paradox especially, Zermelo explains, it seems no longer possible for each logically definable notion to have a set or class as

[41] The mathematical conception is typically identified with the iterative conception of set theory. Quine himself, in *The Roots of Reference* (La Salle, IL: Open Court, 1972, pp. 102–3), recognized this interpretation already in 1972, observing that

> A curiously myopic view of this matter has been manifesting itself of late [fn. omitted]. There is hindsightful reaction, after two generations, to the paradoxes of set theory. The new view is that even before the paradoxes it was not usual to suppose there was a set, or class, for every membership condition. The view is defended by citing Cantor as having already entertained certain restraints on the existence of classes before Burali-Forti published the first of the paradoxes. Fraenkel has undercut this argument by claiming that Cantor already senses paradox [fn. omitted]. What is myopic about the view ... is that it looks back only to the first systematic use in mathematics of the word "set" or "Menge", as if this were uncaused.

[42] Zermelo, "A New Proof of the Possibility of a Well-Ordering," p. 192.

its extension, as had been long thought in the philosophical tradition. Furthermore, Cantor's original definition of a set as "a collection, gathered into a whole of certain well-distinguished objects of our perception or our thought"[43] must also be restricted. What is unsatisfactory about the proposed solutions, including his own, is that they do not preserve these rather simple and intuitive notions of a set. These earlier accounts of the notion of a set have not "been successfully replaced by one that is just as simple and does not give rise to such reservations."[44] Accepting this situation that our intuitions about sets have led us astray, he concedes:

> There is at this point nothing left for us to do but to proceed in the opposite direction and, starting from set theory as it is historically given, to seek out the principles required for establishing the foundations of this mathematical discipline. In solving the problem we must, on the one hand, restrict these principles sufficiently to exclude all contradictions and, on the other, take them sufficiently wide to retain all that is valuable in this theory.[45]

Zermelo here acknowledges – continuing with our ongoing theme – that after the paradoxes, developing set theory must be a largely pragmatic endeavor. The aim in light of the paradoxes is to restrict the notion of set sufficiently to exclude contradiction but to maintain enough of the theory that it can continue its foundational purpose for all mathematics. No solution to such paradoxes will be as intuitive as the notion of a set as the extension of a concept, but this is the situation we must accept if set theory is to go on at all. In this sense, no solution will be entirely satisfactory, including Zermelo's own; there will always be tradeoffs about what to save from our initial intuitive account of sets.

Zermelo then presents his axiom system, attempting to show how he can reduce Cantor's set theory to a few definitions and seven axioms, which appear independent of each other. To indicate that his axioms do in fact preserve a reasonable amount of Cantor's theory, he then develops the theory of cardinal numbers and mentions that he will do the same for the ordinal numbers in a later paper. He admits, however, that he has not been able to give a rigorous proof of their consistency, "though this is very essential." Instead, he will merely indicate how his axioms

43 Cantor, *Beiträge*, p. 84. Georg Cantor, *Contributions to the Founding of the Theory of Transfinite Numbers (1895, 1897)*, trans. Philip E. B. Jourdain (New York, NY: Dover, 1955), p. 85.
44 Ibid., p. 200.
45 Ibid., p. 200.

resolve the known paradoxes in hopes that this will provide a useful beginning toward a consistency proof.[46] Specifically, his separation axiom, he explains, resolves the paradoxes. In contrast to the original comprehension principle, sets can no longer be defined independently by an axiom. Instead, they are always separated out as subsets of sets that have already been given by other axioms. Therefore contradictory notions such as "the set of all sets" or "the set of all ordinals" are excluded from the theory. The defining condition of such subsets must always be definite if the fundamental relations of the domain, meaning the membership relation, by means of the axioms and the logical laws, determine whether or not φ holds for a given value x. He says that in addition to the set-theoretic paradoxes, this last restriction banishes the semantic paradoxes such as the liar paradox and Richard's paradox.[47] It should be noted that this idea of a propositional function being definite would continue to be refined over the next twenty years or so until it would come to mean "definable by means of first-order logic."

Having explained separation, Zermelo next shows in his Theorem 10 how Russell's paradox now becomes the theorem that there is no set of all sets. Every set M possesses at least one subset M_0 that is not an element of M. First, observe that "x is a member of x" is a definite property. Now let M_0 be a subset of M such that x is a member of M_0 if and only if x is not a member of itself. Then M_0 is not a member of M. Suppose M_0 is a member of M_0. Then M_0 is a member of M and M_0 is not a member of M_0 by the construction of M_0. This is equivalent to saying that M_0 is a member of M_0 if and only if M_0 is not a member of M_0, which is a contradiction. Hence M_0 is not a member of M. So the domain itself is not a set, and thus Zermelo's theory eliminates Russell's

[46] Ibid., pp. 200–1. Heinz-Dieter Ebbinghaus, in *Ernst Zermelo: An Approach to His Life and Work* (New York, NY: Springer: 2007, pp. 77–8), criticizes Moore's interpretation from the side of consistency, observing that this was crucial to Hilbert's program in the philosophy of mathematics. Thus, much as I have been urging, Zermelo's axiomatization is not a purely mathematical undertaking.

[47] Zermelo, "Investigations into the Foundations of Set Theory I," pp. 201–4. The liar paradox is probably familiar to most readers and comes in a variety of forms. One of the simplest tells of the person who asserts, "I am lying." If this is a lie, then it is not. And if this is not a lie, then it is. Richard's paradox is more complicated but relies on the notion of definability and a diagonal method of the sort found in Cantor's proof that the set of real numbers is uncountable. If we let X be the set of all definable real numbers, we have a countable set, since there are only countably many definitions. If we order these definitions lexicographically, that is, in dictionary order, and then apply the method of diagonalization to this ordering, we get a new real number that is not in our set of definable real numbers. But have we not just defined this number by way of this procedure?

paradox, turning it instead into a theorem.[48] Let us now consider Russell's own resolution to the paradoxes.

Already in his *Principles of Mathematics*, Russell proposes solving the paradoxes with his theory of types, a sort hierarchy of propositional functions that would prevent such membership conditions as self-membership and hence would exclude from the theory his own paradox, among others.[49] However, Russell's proposal of types in the *Principles* is tentative, and between 1903 and 1908, he explored a number of other approaches to resolving the paradoxes before coming back to types in his 1908 "Mathematical Logic as Based on the Theory of Types."[50] He begins here by remarking that this system of logic recommended itself to him because of its ability to resolve certain paradoxes, in particular Burali-Forti's paradox of the largest ordinal number. "But the theory in question seems not wholly dependent on this indirect recommendation," he argues, saying that "it has also, if I am not mistaken, a certain consonance with common sense which makes it inherently credible. This, however, is not a merit upon which much stress should be laid; for *common sense is far more fallible than it likes to believe*" [my emphasis].[51]

We should note, here, how similar Russell's approach is to Zermelo's, both in starting from the paradoxes and in motivating their proposed solution by the success of resolving them. Neither Russell nor Zermelo relies on some preconceived notion of set beyond just the mathematical aims of the theory. Beyond this, there is no univocal conception of a set that their respective renderings of the theory aim to capture. They each recognize that it was our intuitions about sets that led the theory astray, so these can no longer serve as the primary guide in further developing the theory. For both Russell and Zermelo, what determines the success of their axiomatizations is their ability to avoid the paradoxes while still including enough of Cantor's original theory to serve as a foundation for mathematics. Again, like Zermelo, Russell takes Cantor's attitude of developing the theory in response to mathematical concerns and applies it to the resolution of the paradoxes themselves.

48 Zermelo, "Investigations into the Foundations of Set Theory I," p. 203.

49 Russell, *Principles of Mathematics*, pp. 103–5 and Appendix B.

50 This is particularly true of Russell's 1905 "On Some Difficulties in the Theory of Transfinite Ordinals and Order Types," in *Essays in Analysis*, ed. Lackey (London: George Allen and Unwin, 1973), pp. 135–64. I will come back to this essay in discussing Quine's Russellian background.

51 Bertrand Russell, "Mathematical Logic as Based on the Theory of Types," in *From Frege to Gödel*, ed. van Heijenoort, p. 153. This view, as we will see in Chapter 4, is strikingly similar to Quine's statement that in light of the paradoxes, "common sense is bankrupt."

Unlike Zermelo, Russell next gives a detailed account of the various paradoxes, among them the liar paradox, the Burali-Forti paradox, and his own.[52] Russell's detailed probing of the various paradoxes aims to discover that they all share "a common characteristic."[53] The common feature that Russell discerns is "self-reference, or reflexiveness." So, for example, when Epimenides the Cretan asserts that all Cretans are liars and that all other statements made by them are lies, this assertion itself must fall within the scope of Epimenides' statement. Similarly, for all classes x to belong to the class w if and only if class x is non-self-membered, this condition must apply to itself. So, Russell explains, "[i]n each contradiction something is said about *all* cases of some kind, and from what is said a new case seems to be generated, which both is and is not of the same kind as the cases of which *all* were concerned in what said" [Russell's italics].[54] He then proceeds through the paradoxes, showing how this works in each case and concluding, "Thus all our contradictions have in common the assumption of a totality such that, if it were legitimate, it would at once be enlarged by new members defined in terms of itself."[55] Thus Russell establishes the general principle – a version of his vicious circle principle – that "[w]hatever involves all of a collection must not be one of the collection" or, conversely, "[i]f, provided a certain collection had a total, it would have members only definable in terms of that total, then the said collection has no total."[56]

The theory of types will be a theory that excludes such collections, making such phrases as "all propositions," "all properties," or "all classes" meaningless. I should make clear at the outset of this discussion that Russell's theory is not the perhaps familiar modern type theory, now known as the simple theory of types. Russell's theory is what is called the ramified theory of types. I will make the distinction clear, but let me begin with Russell's theory as he presents it in 1908.

Russell begins by defining a type as the range of significance of a propositional function, that is, the collection of arguments for which the function has values or of which it is true or false. If a propositional function is applied to an argument outside of its range of significance, the resulting proposition is not a proposition at all but rather is simply

[52] Zermelo, "Investigations into the Foundations of Set Theory I," pp, 200–1.

[53] Russell, "Mathematical Logic as Based on the Theory of Types," p. 154.

[54] Ibid., p. 154.

[55] Ibid., p. 155.

[56] Ibid. For a more detailed account of this principle and its motivations in Russell, see Peter Hylton, "The Vicious Circle Principle," in *Propositions, Functions, and Analysis: Selected Essays on Russell's Philosophy* (Oxford: Clarendon Press, 2005), pp. 108–14.

nonsense. Whenever an apparent variable, that is, a bound variable, occurs in a proposition, its range of values forms a type. The paradoxes then are avoided by the vicious circle principle as implemented by the theory of types: "No totality can contain members defined in terms of itself" or, in terms of variables, "whatever contains an apparent variable must not be a possible value of that variable." Russell's typing of the objects of the universe restricts it so that any expression containing an apparent variable must be of higher type than the possible values of that variable. Thus the apparent variables in an expression determine the type of the expression.[57] He explains further that any proposition containing an apparent variable is a "generalized proposition," and any proposition with no apparent variables is an "elementary proposition." A generalized proposition presupposes the other propositions from which it is obtained by the process of generalization. Hence all generalized propositions presuppose elementary propositions. In an elementary proposition, we distinguish one or more terms from one or more concepts, where the terms are what we regard as the subject of the proposition and the concepts are the predicates or relations asserted of the terms. Russell calls the terms of an elementary proposition the "individuals," and these form the lowest type. In practice, however, he states that it is unnecessary to know which objects belong to the lowest type or even to know, in a given context, whether the lowest type are individuals. The lowest type in any context, then, may be taken as the individuals, and all that is essential to them is the way in which the other types are generated from them. All that ultimately matters are the relative types of the objects in a given expression – a view that Russell describes as "typical ambiguity" and a key insight to which we will see W. V. Quine appealing in developing his own set theory.[58]

With all of this in place, Russell's theory then generates the type hierarchy as follows. In generalizing the individuals in an elementary proposition, we generate a new proposition, and this will be a legitimate procedure as long as no individuals are propositions, which is the case because propositions are essentially complex whereas individuals lack all complexity. So, for example, given the proposition "x is red," where "x" is taken as an individual variable, we may generalize on "x" to get the new proposition "$(\forall x)(x \text{ is red})$," or "Every individual is red." Since the individuals cannot be propositions, generalizing on them in this way does not lead to reflexive paradoxes. Russell calls the elementary

[57] Russell, "Mathematical Logic as Based on the Theory of Types," pp. 162–3.
[58] Ibid., pp. 163–4.

propositions along with the propositions containing only individual variables as apparent variables "first-order propositions," and these propositions form a new set of objects that can also be generalized so as to generate the second type. Generalizing now on the variables of the second type, we generate the second-order propositions, which now also serve as quantifiable variables so as to form the third type, and so on. Now we can see how the typing removes the reflexive paradoxes. Epimenides' statement now becomes, for example, "All first-order propositions asserted by me are false." Since this is a proposition about first-order propositions, it is itself a second-order proposition and so may be asserted truly. It does not assert any true first-order proposition, and so does not yield a paradox as it previously did. In general, the $(n + 1)$th logical type consists of order n propositions, and these propositions contain as apparent variables propositions of order no higher than $n - 1$. Since the types are mutually exclusive, it is impossible to generate the reflexive paradoxes.[59]

In practice, Russell explains that it is easier to work with a hierarchy of functions than with one of propositions. Thus a function that takes only individuals as arguments, thus always yielding a first-order proposition, is a first-order function. A function involving only first-order functions or propositions as apparent variables is a second-order function and so on. Furthermore, he calls a function of one variable that is of order next above its argument a "predicative function"; this is the same for propositions of several variables. Hence the type of a function is determined by the type of its values and the type and number of its arguments. Russell further explains this hierarchy by denoting first-order functions of an individual variable x as "φ!x." Since no first-order function contains a function as an apparent variable, hence forming a well-defined totality, we can generalize on φ. Then any proposition with "φ" as apparent variable and no apparent variable of higher type is a second-order proposition. If such a proposition contains an individual variable x, then this proposition is not a predicative function of x but it is a predicative function of φ, where φ is written "f!(ψ!ŵ)." So f is a predicative function, and its values form a well-defined totality, so we can also generalize on f. This defines the third-order predicative functions, which are functions that have third-order propositions as values and second-order predicative functions as arguments and so on.[60]

Let me step back a bit from Russell's exposition here to make a bit clearer how the ramified type theory works, in particular by clarifying the

[59] Ibid., p. 164.
[60] Ibid., p. 165.

relationship between types and orders. At this point, it will be useful to introduce the perhaps more familiar modern simple theory of types, most often attributed to F. P. Ramsey but discovered independently by Quine as well in his 1932 dissertation, "The Logic of Sequences."[61] Since Russell always leaves his variables typically ambiguous, we will make some assumptions here not found in his exposition. The simple type hierarchy may be described as follows. At the lowest type are the individuals; these we assign type 0. Propositional functions that apply to individuals are of type 1. And then functions that apply to these are of type 2 and so on. The types are further distinguished according to how many arguments the functions involved take. Since a propositional function of level 1 applies only to an individual, self-reflexivity is ruled out here, since such a propositional function can never apply to itself without violating the type restrictions. Clearly, such paradoxes as Russell's are then blocked if we treat this theory extensionally, that is, treat the propositional functions as classes and predication as membership. Then no class can be a member of itself.[62]

Recall, however, that Russell's vicious circle principle was implemented by way of restricting the range of the quantifiers, as encapsulated in the form "whatever contains an apparent variable must not be a possible value of that variable." Here is where the theory of orders enters. The simple type theory blocks the paradoxes of set theory by way of ruling out self-membership; we just saw an example of this. The account is slightly more complicated, but for similar reasons, Cantor's paradox and the Burali-Forti paradox are also ruled out. But as we also saw, Russell initially shows instead how his theory blocks one of the semantic paradoxes: the liar paradox (he later goes on to discuss how the set-theoretic paradoxes are blocked by his ramified theory).[63] The theory of orders is aimed at these. In the liar paradox, we violate the vicious circle principle

[61] F. P. Ramsey, *The Foundations of Mathematics and Other Essays*, Ed. R. B. Braithewaite (New York: Humanities Press, 1950), pp. 1–61.

[62] More detailed accounts of the simple theory of types are abundant in the literature; see, for example, A. A. Fraenkel, Y. Bar-Hillel, and A. Levy, *Foundations of Set Theory*, 2nd rev. edn (Amsterdam: North-Holland, 1973), pp. 158–61.

[63] Russell did not distinguish between the set-theoretic and semantic paradoxes. He thought that a successful resolution to the paradoxes would block both sorts. This is not an unreasonable demand, given that his starting point was an intensional theory of propositional functions. I will discuss this a bit more in Chapter 4. Ramsey is usually given credit for distinguishing the semantic from the set-theoretic paradoxes, in the same paper in which he proposed his simple theory of types. The simple theory, too, prevents the semantic paradoxes but by way of keeping metalinguistic claims concerning truth, definition, denotation, meaning, and the like distinct from the object level claims that are assertible from within the system.

by asserting a proposition, which would then itself fall into the range of the apparent variable within that assertion. "All propositions asserted by me are false" is itself a proposition and so would be a value of the apparent variable used in talk of all propositions. The theory of orders prevents such enlargement of the totality of propositions by prohibiting such use of quantification. This is not the same as the simple theory of types.

To see this, consider the following two propositions:

(1) Napoleon is brave.
(2) Napoleon has all the qualities of a great general.

In the simple theory of types, the propositional functions "û is brave" and "û has all the qualities of a great general" are of the same logical type, since they both hold of individuals, assuming that people are among the individuals. (Russell does not assume this.) In the ramified theory, though, they both hold of individuals but are of different order, since (2) would involve quantification over all the particular qualities, such as bravery and leadership, that a great general has. Finally, we should also distinguish Russell's theory of orders from modern usage in first-, second-, third-, and higher-order logics generally. The modern use of these terms would allow quantification only over individuals, over first-order functions, over second-order functions, and so on. Russell's theory allows for further possibilities. For example, functions of functions of individuals are of second order, but so are functions such as "û has all the qualities of a great general," since these, too, involve quantification over the totality of first-order functions.[64]

We saw earlier with the example of Epimenides how Russell's type theory blocks at least some of the paradoxes, but now a new problem arises. Recall that the success of his theory was to be judged on its capacity for providing a foundation to mathematics. As it stands, it generates very little of mathematics. The typing restrictions we saw exclude such expressions as "all propositions" or "all properties of x"; we can make such assertions only when restricted to a particular order. But if a finite number, for example, is defined in the usual way as a number that possesses all properties possessed by zero and by successors of all numbers

[64] I have benefited in my exposition here from Alonzo Church's "Comparison of Russell's Resolution of the Semantical Antinomies with That of Tarski," *Journal of Symbolic Logic* 41:4 (December 1976), pp. 747–60, and from Bernard Linsky's "The Notation in *Principia Mathematica*," https://plato.stanford.edu/entries/pm-notation/. Linsky's exposition largely follows Church's but is less technical in nature and more attentive to the details of Russell's actual presentation.

possessing these properties, then type theory prohibits such a definition, since it says we must confine our discussion to properties of a particular order. If we confine this statement to all first-order properties of numbers, then we cannot infer that this definition holds for all second-order properties. So, for example, we cannot prove that if m and n are finite numbers, then $m + n$ is a finite number, since by the previous definition, "m is a finite number" is a second-order property of m. So from "$m + 0$ is finite, and if $m + n$ is finite, so is $m + n + 1$," we cannot conclude by induction "$m + n$ is a finite number," leaving much of mathematics impossible.[65] Russell answers this difficulty by introducing his axiom of reducibility.

To regain mathematics, Russell needs a way of reducing the order of a propositional function without affecting the truth or falsity of its values, thus allowing him to simulate talk of all properties. To do so, he adopts what he takes to be the commonsense reason for accepting classes. He explains that, given any propositional function of any order, φx, we assume the function to be equivalent, for all arguments x, to the statement "x belongs to the class α." This is a first-order statement, which makes no mention of all functions of such-and-such a type, and its only practical advantage over the original statement is its being of first order. Russell sees no particular advantage in assuming outright the reality of classes (instead defining classes contextually from his intensional propositional functions):

> I believe the chief purpose which classes serve, the chief reason which makes them linguistically convenient, is that they provide a method of reducing the order of a propositional function. I shall, therefore, not assume anything of what may seem to be involved in the common-sense admission of classes, except this: that every propositional function is equivalent, for all its values, to some predicative function.[66]

Much as with Zermelo's system and the basic system of types, the axiom is introduced on purely pragmatic grounds. Russell, too, aims to preserve as much of set theory as mathematics requires while still excluding the paradoxes. Beyond this goal, there is no prior conception of a set for which he aims. After all – as he was all too aware – the paradoxes showed that neither common sense nor our intuitions about sets were reliable guides in developing the theory.

Russell elaborates the new axiom, explaining that it is to hold for any function, regardless of the type of its argument. So if we let φx be a function of any order of an argument x, which itself may be of any type,

[65] Russell, "Mathematical Logic as Based on the Theory of Types," p. 167.
[66] Ibid.

then by the axiom of reducibility, there is a predicative function $\varphi!x$ that is extensionally equivalent to it. This, Russell thinks, is all the usual assumption of classes is meant to do and is as much of classes as he is willing to assume. It is all he needs for mathematics. The axiom allows statements about "all first-order functions" or "all predicative functions" to do the work that statements about "all functions" previously did. "The essential point," he observes, "is that such results are obtained in all cases where only the truth or falsehood of values of the functions concerned are relevant, as is invariably the case in mathematics."[67] In other words, mathematics requires only extensional equivalence. Yet there may be a worry that with the axiom, the paradoxes return. Russell argues that this is not the case, since in the case of the paradoxes, either something beyond the truth or falsity of the values of the function is concerned or the expression occurs that remains unmeaningful even after the axiom is applied, as he then goes on to show, with regard to Cantor's paradox and the Burali-Forti paradox.[68]

Russell closes his paper much as Zermelo began:

> The theory of types raises a number of difficult philosophical questions concerning its interpretation. Such questions are, however, essentially separable from the mathematical development of the theory and, like all philosophical questions, introduce elements of uncertainty which do not belong to the theory itself. It seemed better, therefore, to state the theory without reference to philosophical questions, leaving these to be dealt with independently.[69]

It seems then that where set theory is concerned, the line between philosophy and mathematics remains a difficult one to draw. But for Cantor, Zermelo, and Russell, what must be preserved in set theory is its ability to recapture ordinary mathematics, and they do so, guided by just this pragmatic concern. We turn now to Quine and his set theory New Foundations (NF), in which we will see this attitude toward set theory developed more explicitly as a general approach to the subject.

67 Ibid., p. 168.

68 Ibid., pp. 168, 179–82. For an excellent technical exposition of this point with regard to the heterological paradox and of the theory of types in general, see Church, "Comparison of Russell's Resolution of the Semantical Antinomies with That of Tarski," pp. 747–60, especially pp. 752–4.

69 Russell, "Mathematical Logic as Based on the Theory of Types," p. 182. It should be noted that Moore attributes the discovery of Russell's paradox to the important intertwining of mathematics and philosophy. In "The Origins of Russell's Paradox," pp. 23–45. Moore remarks, "Mathematics and philosophy have interacted in many ways, but in the twentieth century mathematicians have often reacted with suspicion when philosophers had something to say about mathematics. The origins of Russell's Paradox provide a case study of how traditional philosophical concerns led to genuine mathematical progress."

CHAPTER 3

New Foundations and the Beginnings of Quine's Philosophy of Set Theory

We now turn to consider the system of set theory put forward by W. V. Quine in his 1937 paper "New Foundations for Mathematical Logic."[1] New Foundations (NF), as the system itself is known, emerged from Quine's thinking through the systems of both Bertrand Russell and Ernst Zermelo, preserving something like the type restrictions of the one theory along with the unrestricted variables of the other. Furthermore, Quine adopts the preservation of set theory's mathematical contributions – most explicitly of Georg Cantor, Russell, and Zermelo – as the criterion of success for his system. NF is especially notable – and controversial – in providing for the existence of such "big" sets as the set of ordinals, the set of cardinals, and even the universal set. In doing so, it makes mathematical sense out of Cantor's absolute infinities, or supposedly inconsistent totalities, and shows that Russell was not entirely unjustified in his earlier commitments to a class of all classes. In this chapter, I lay out the system and then, as in Chapter 2, discuss how it avoids the paradoxes. This raises the issue of the consistency of Quine's system as well as consistencies for set theories generally. I conclude with some remarks from a Quinean standpoint on the mathematical and philosophical significance of consistency proofs. I also begin to examine, from the perspective of Quine's work in logic and foundations, his more general philosophical approach to set theory, which will be the primary focus of Chapters 4 and 5.

3.1 From Russell and Zermelo to New Foundations

Quine begins by observing that Bertrand Russell and Alfred North Whitehead's *Principia Mathematica* provides substantial evidence that all of mathematics is translatable into logic, although in light of Kurt

[1] W. V. Quine, "New Foundations for Mathematical Logic," *The American Mathematical Monthly* 44:2 (1937), pp. 70–80; reprinted in an expanded form in *From a Logical Point of View*, pp. 80–101.

Gödel's incompleteness theorem, no axiom system is sufficient for deducing all of mathematics. Still, the core of mathematics, as in *Principia Mathematica*, serves as a reasonable standard for judging any system to be sufficient for mathematics, so it is this standard that Quine adopts for judging NF. Like Zermelo and Russell, Quine judges the success of any given formulation of set theory according to whether or not the system captures some reasonable amount of mathematics. As for logic, Quine admits that the logic required for mathematics is a good deal stronger than traditional logic:

> It must be admitted that the logic which generates all this is a more powerful engine than the one provided by Aristotle. The foundations of the Principia are obscured by the notion of propositional function, but, if we suppress these functions in favor of the classes and relations which they parallel, we find a three-fold logic of propositions, classes, and relations. The primitive notion in terms of which these calculi are ultimately expressed are not standard notions of traditional logic; still they are of a kind which one would not hesitate to classify as logical.[2]

So we have Quine clarifying exactly what he will count as a logical foundation for mathematics: logic inclusive of set theory. But in accepting such a position, he knowingly gives up much of the epistemological aims that the reduction of mathematics to logic was supposed to accomplish. Quine's concern is not to somehow rest mathematics on a firmer foundation than mathematics itself. Rather, this foundation will be internal to mathematics itself. However, for Quine, set theory, as we will see later, remains foundational in that it unifies diverse bodies of mathematics into the single framework of logic and set theory.[3] This theme is important for understanding Quine's philosophy of set theory as a whole, which we will deal with more thoroughly in the following chapters. For now, we merely observe that these departures from traditional philosophical aims are present in some of his earliest work in foundations. However, this is not the only point of interest here in beginning to flesh out Quine's philosophical stance. We also see him in this passage quickly dismissing Russell's basic notion of a propositional function on grounds of its

[2] Ibid., p. 71.

[3] This point is also interesting in light of later disputes over the logical status of higher-order logic. We see here that Quine was already in a position at this early stage to draw a distinction between first- and higher-order logic on grounds of their respective claims to logicality. In a way, what we see here is that the label "logic" was never of great importance to Quine. What was important was making clear what it was we were willing to count as logic and being honest about what sorts of assumptions we are building into attributing the label "logic." Among such assumptions, as in the case here, is the mathematical strength of such theories.

obscurity and instead accepting classes outright. The central Quinean themes of clarity and simplicity are present already at this early stage, and in recognizing the obscurity of the notion of a propositional function, we can see the beginnings of Quine's attack on philosophical accounts of meaning and so also on the analytic/synthetic distinction. We see here core concerns of Quine's philosophy emerging from his logical work. Let us turn now to the system itself.

The system of NF contains a much smaller number of basic notions than that of *Principia Mathematica*, employing just membership, "$\in$" (epsilon); alternative denial, "|" (not both); universal quantification, "$(\forall x)$"; and an infinite number of variables such as "x," "y," "z," "x'," "y'," "z'," "x''," and so on. The variables in NF, unlike those of *Principia Mathematica*, range over all objects whatsoever; there is no stratification of the universe into different levels or types. From these basic notions, Quine then defines the usual logical connectives of negation, conjunction, disjunction, conditional, biconditional, and the existential quantifier, along with the various set-theoretic relations such as subset, "$\subseteq$"; union, "$\cup$"; and intersection, "$\cap$." In remarking that the formula "$(x \in y)$," or "x is a member of y," makes sense only where y is a class, Quine further simplifies the system by adopting the convention that where y is an individual, we may interpret such formulas as "x is the individual y." As such, every individual x is just equal to its unit class, or singleton, $\{x\}$, but Quine adds in a footnote that "this is harmless."[4] Again, we see the broadly pragmatic spirit that has come to characterize Quine's philosophy emerging in his logical work. Where he can simplify the system, unlike – and in reaction to – Russell, he does so without regret.

Next, Quine introduces the basic rules and axioms, or initial theorems, of NF, among them truth-functional and quantificational axioms and rules of inference. He also proposes the following two set-theoretic principles:

P1. $(\forall x)(\forall y)(x \subseteq y \supset (y \subseteq x \supset x = y))$,

the extensionality principle that a class is determined by its members, and

R3. If "x" does not occur in φ, then "$(\exists x)(\forall y)((y \in x) \equiv \varphi)$" is a theorem.

[4] Quine, "New Foundations for Mathematical Logic," p. 71. This convention is indeed harmless, as shown by Scott, but it also results in the surprisingly very different system of NFU, which was proved consistent in Peano arithmetic by Jensen.

In short, given any condition on y, there is a class x with members y such that this condition holds for the y's. Quine recognizes that this is, of course, the inconsistent comprehension principle, which yields Russell's paradox when we take φ as the condition "$(y \notin y)$." As we saw in Chapter 2, one way of resolving the paradoxes was to adopt types so that the universe is stratified into levels and the variables are restricted to ranging over objects of a particular level. Thus $(\alpha \in \beta)$ is a formula only if the values of β are of type $n + 1$ and the values of α are of type n. In all other cases, $(\alpha \in \beta)$ is neither true nor false but meaningless.[5]

Quine, however, focuses his attention on an observation that we saw Russell himself make in the previous chapter:

> In all contexts the types appropriate to the several variables are actually left unspecified; the context remains systematically ambiguous, in the sense that the types of its variables may be construed in any fashion conformable to the requirement that "∈" connect variables only of consecutively ascending types. An expression which would be a formula under our original scheme will hence be rejected as meaningless by the theory of types only if there is no way whatever of so assigning types to the variables as to conform to this requirement on "∈." Thus a formula in our original sense of the term will survive the theory of types if it is possible to put numerals for the variables in such a way that "∈" comes to occur only in contexts of the form "$n \in n + 1$."[6]

While Russell had already noticed this feature of types, Quine follows it through to its full conclusion. It is inessential to resolving the paradoxes that the universe actually be stratified into levels. What the typing restriction does is to provide a syntactic test for which formulas actually determine classes; these are the so-called stratified formulas. NF, in this sense, as Quine later remarked, emerged from the theory of types.[7] Here, we get early hints of the philosophical attitude he adopts toward set theory. His philosophical vision is served if he can somehow simplify the theory and, furthermore, avoid some of the metaphysical assumptions about set theory that Russell's hierarchy of types has often inspired.

To help motivate this move, Quine notes some unnatural consequences of Russell's type theory. Although we saw Russell clinging desperately to a universal class, his resolution of the paradoxes in type theory ultimately forced him to give up this idea in favor of a series of what

[5] Ibid., p. 77.
[6] Ibid., p. 78.
[7] W. V. Quine, "The Inception of New Foundations," in *Selected Logic Papers*, enlarged edn (Cambridge, MA: Harvard University Press, 1995), p. 287.

Quine describes as quasi-universal classes, a series of classes of all objects of the type directly below a given type. So there is a class of all individuals, a class of all classes of individuals, and so on, but there is no single class V that contains all objects regardless of type. Similarly, there is no single null class Λ but, again, one for each type. Type theory has the further odd consequence that classes no longer have absolute complements. A class $-x$ does not contain all nonmembers of x but rather only those nonmembers of x at the next lower type from x. So we see that the Boolean class algebra no longer applies to classes in general but only to classes within particular types. Furthermore, these classes are then reduplicated within each type, and the same holds also for the usual calculus of relations. Finally, Quine notes that even arithmetic is reduplicated in each type so that the numbers fail to be unique, there being for each type a number 0, 1, 2, 3, and so on. He concludes, "Not only are all these cleavages and reduplications intuitively repugnant, but they call continually for more or less elaborate technical maneuvers by way of restoring severed connections."[8] By taking Russell's insight of typical ambiguity seriously, Quine resolves the paradoxes while avoiding these unhappy consequences: "Whereas the theory of types avoids the contradictions by excluding the unstratified formulas from the language all together, we might gain the same end by continuing to countenance unstratified formulas but simply limiting R3 explicitly to stratified formulas."[9] The hierarchy of types is then removed so that variables stay unrestricted in range and the logical language allows for all formulas in the original untyped sense. The notion of a stratified formula remains a part of the system only in the following revised version of R3:

R3′. If φ is stratified and does not contain "x," then "$(\exists x)(y)((y \in x) \equiv \varphi)$" is a theorem.

Here, we also get a glimpse of a further reason for rejecting the theory of types that Quine was to emphasize in his later account of how he came upon NF. There, he explained that he disliked the arbitrary grammatical restrictions of type theory. As was mentioned previously, unstratified formulas, according to type theory, were neither true nor false but meaningless, so "[s]eemingly intelligible combinations of signs were banned as ungrammatical and meaningless."[10] The theory of types was,

8 Ibid., pp. 78–9.
9 Ibid., p. 79.
10 Quine, "Inception," p. 286.

for Quine, too drastic in its resolution of the paradoxes in banning all unstratified formulas, especially since many of them seem perfectly intelligible. Much of his work up to NF was an attempt to legitimize these restrictions so that the banning of unstratified formulas would not be wholesale and without reason.[11] In his first attempt in 1935, he tried to legitimize the restrictions by finding some more basic array of notions that would allow him to provide contextual definitions of classes and the membership relation. "I hoped," he explained, "to devise contextual definitions that would generate just the formulas that fit the theory of types, while leaving other formulas meaningless in the straightforward sense of not being accounted for by the definitions." The attempt failed, however, as the set-theoretic paradoxes simply shifted to become semantic paradoxes.[12]

At this point, the crucial influence on Quine shifted from Russell to Zermelo. From Zermelo, Quine gained the insight "that a meaningful open sentence may or may not determine a class, and that it can be left to the axioms to settle which ones do."[13] So while Russell banned such open sentences as "x is a member of itself" on the grounds of their supposed meaninglessness, Quine took the contrary position, affirming their meaningfulness but leaving it open that they may not determine a class, depending on which classes the axioms yield. His first step in the direction of NF was the 1936 "Set-theoretic Foundations for Logic," in which he presents a version of Zermelo's system but with a modified version of the separation axiom. Zermelo states his original version of separation as

$$(\exists x)(\forall y)(y \in x \leftrightarrow (x \in z \wedge Fy)).$$

Quine's idea was to replace "$x \in z$" with "$x \subseteq z$" so as to obtain

$$(\exists x)(\forall y)(y \in x \leftrightarrow (x \subseteq z \wedge Fy)).$$

While Zermelo's system required several other set-theoretic axioms to be adequate for the derivation of mathematics, Quine's modified system, which he called "Γ," included only the modified separation axiom and the extensionality axiom. In line with the motivations for NF, his aim in "Set-theoretic Foundations" was "the presentation of a system Γ which resembles [Zermelo's] system but is more economical."[14] Like the systems of Zermelo and Russell, however, to derive a substantial amount of mathematics (namely, all of what could be derived in the system of

[11] Ibid.

[12] Ibid., pp. 286–7.

[13] Ibid., p. 288.

[14] W. V. Quine, "Set-Theoretic Foundations for Logic," in *Selected Logic Papers*, p. 85.

Principia Mathematica), the system Γ still required postulating the additional axioms of infinity and choice.

Indeed, this lack of economy was the primary drawback Quine saw in Zermelo's system, since it otherwise avoided the difficulties of types in that it lacked both arbitrary grammatical restrictions and the reduplication of objects at each level of the hierarchy: "Zermelo's system itself was free of both drawbacks, but in its multiplicity of axioms it seemed inelegant, artificial, and ad hoc."[15] But both Zermelo's system and the modified version Γ had another drawback in removing type theory's reduplication of objects. While types had quasi-universal classes, Zermelo's system banned this class altogether; therefore, absolute complements were again impossible, yielding instead only complements with regard to an already given class. In general, what both systems lacked were big classes.[16] Here, we see another aspect of Quine's philosophy of set theory, which will emerge more fully in Chapter 4. Quine approaches set theory as an explorative project much like the rest of science. The lack of big classes is not necessarily a drawback for reconstructing mathematics within set theory. Most ordinary mathematics goes on with comparatively small sets, but such a lack becomes all the more apparent as set theory itself has emerged as a subject of mathematical investigation.[17] In exploring the mathematical behavior of the set-theoretic universe, we should not limit our explorations to only some sets. We should want to know what all sets are like, how they are similar and where they diverge, and what sorts of features cause such divergences.[18] If we want to know what the big sets are like, neither the system of Russell nor that of Zermelo can give us an answer.

What Quine wanted was a system that would be free of all the drawbacks of both systems "but would be like the theory of types in having a single comprehension principle of class existence, and would admit big

[15] In later chapters, we will see more of how Quine revised this view of Zermelo's set theory in light of the connections he was able to draw between it and type theory. Drawing out these sorts of connections is central to Quine's philosophy of set theory and to his philosophy in general. He gestures at this move in "The Inception of New Foundations," pp. 287–8. Contrast this with his earlier statement in "New Foundations for Mathematical Logic" (p. 80, fn.) about the awkwardness of Zermelo's system, which he notably removed from the paper in the form published in *From a Logical Point of View*.

[16] Quine, "The Inception of New Foundations," p. 288.

[17] For this history, see Kanamori's "Development," pp. 1–71.

[18] This is perhaps very much in the spirit of much current research in set theory that is concerned with large cardinal axioms. We want to know what the universe is like beyond the sets given by ZFC. Yet NF remains largely neglected in its possible insights into the universe.

classes without restriction to type."[19] Combining the insights of Russell's typical ambiguity with Zermelo's view that the axioms could determine which sets exist gave Quine precisely this, as codified in his system NF. In light of Zermelo, "I was able to look to types as a restriction specifically upon classes and not upon language," Quine observed. "The purpose of the theory of types was to bar the paradoxes, and this could be done by using it only to say which open sentences are to be taken to determine classes."[20] The structure of the set-theoretic universe, then, was determined by the stratification condition limited only to the comprehension principle R3′. The universe itself was not stratified into a hierarchy of classes, as Russell thought. Now, with the variables reconstrued as general, that is, as ranging over all objects with no restriction of type, Quine "found new strength accruing at every turn, apparently with impunity."[21] NF successfully removed Russell's arbitrary grammatical restrictions while not adding the reduplication of objects back in. Furthermore, he regained the universal class and absolute complements as well as other such big classes as the class of cardinals and the class of ordinals. Finally, he obtained an infinite class without the arbitrary postulation of an axiom of infinity in that the universal class existed and contained within in it each natural number such that for all such numbers m and n, $m \neq n$.[22]

Quine's aim in presenting NF was not to do away with other set theories, to have somehow presented the single "right" conception of a set. Rather his aim was, and would remain, to further set-theoretic research generally by looking at sets from a different perspective, which preserved certain fairly intuitive features of sets that the systems of Russell and Zermelo did not. From Quine's perspective, none of these systems discovered anything like the essence of sets so as to discover the one and only correct set theory. Rather, just like Russell's and Zermelo's, Quine's primary purpose was eliminating the paradoxes while preserving enough of Cantor's original theory for it to be mathematically interesting. Any set theory that was adequate to these aims was one worthy of mathematical – or perhaps we might even say more broadly – scientific investigation. As we will see in Chapter 4, Quine's philosophy of set theory can be encapsulated in the notion of explication as opposed to conceptual analysis.

19 Quine, "The Inception of New Foundations," p. 288.
20 Ibid.
21 Ibid.
22 However, it would not be until 1953 that Ernst Specker showed outright that the axiom of infinity was provable in NF.

3.2 "Contradictions Really Scare Me": New Foundations and the Paradoxes

But did NF in fact avoid the paradoxes?[23] Quine remarks in conclusion that he has no proof that such a system is consistent, but just as Zermelo and Russell reasoned about their respective systems, Quine can see no way of deriving a paradox within it. He himself makes this connection: "The lack of a consistency proof is no special ground for misgivings, for there is likewise none for the systematization involving the theory of types."[24] Furthermore, his system removes the various unnatural consequences of types. Both the universal class V and the null class Λ are unique, while the complement of *x*, –*x*, is the class of everything not in *x*. Thus the Boolean class algebra is restored and similarly for the calculus of relations. The reduplications of arithmetic are also removed so that the numbers are unique and their laws become generally applicable as a single calculus, thus eliminating whatever complicated technical devices type theory required for restoring such severed connections.[25]

In concluding his paper, Quine remarks that NF differs from the original inconsistent theory in preventing the existence of classes defined by unstratified conditions. He adds in a footnote that the systems of his earlier "Set-theoretic Foundations for Logic" and Zermelo's offer other approaches for resolving the paradoxes. "But these methods entail most of the awkward limitations which are entailed by the theory of types," he argues. "The present method of avoiding the contradictions, if it indeed avoids them, would seem to be the least restrictive method yet suggested."[26] So we see again that Quine's guideline for developing set theory is much like Zermelo's and Russell's: to preserve as much as the original theory unrestricted theory as possible. In this sense, NF does seem an advance over Zermelo's system and theory of types. While both of these versions of set theory are sufficient for reconstructing mathematics, they lack certain sets, the so-called big sets in particular. As we have seen, these other theories include a variety of inelegancies resulting from the way in which they restrict the notion of set so as to avoid the paradoxes.

So far, we have had only Quine's speculations that NF is free from contradiction. It does seem fairly obvious that Russell's paradox will not

[23] The quote in the section title comes from my friend, Jon Snodgrass; see his "History with History," from the album *Drag the River* (Last Chance Records, 2013).
[24] Quine, "New Foundations for Mathematical Logic," p. 79.
[25] Ibid.
[26] Ibid., p. 80.

be a source of contradiction in the system, given that there is no set with the membership condition $y \notin y$.[27] But how exactly the theory avoids the other paradoxes, such as Cantor and Burali-Forti, is less clear. Furthermore, as we saw in Chapter 1, Cantor had arrived at an important result of set theory – the theorem bearing his name – that the power set of a set always has greater cardinality than the set itself. It seems that we should want to preserve such a result if we are to preserve Cantor's set theory. I now examine how in fact the known paradoxes are blocked in NF and also how the theory preserves an analog of Cantor's theorem.

Recall from Chapter 2 that Cantor's paradox results from the claim that there exists a universal set and Cantor's result that for each set x, the power set of x has greater cardinality than the set x itself. Quine's NF certainly yields a universal set V in its stratified existence claim that $(\exists x)(\forall y)(y \in x \equiv y = y)$.[28] So it seems that such a set should have the greatest cardinality, given that it contains all sets. But then by Cantor's theorem, the cardinality of the power set of V would be greater than V itself, landing us back in the contradiction that worried Russell so much. How then does NF block Cantor's paradox? In the first general result about NF, Quine himself showed how his system both blocks this paradox and preserves an analog of Cantor's theorem.[29] Although it may seem tedious, it is instructive to follow the details of this proof to see exactly how stratified comprehension works in preventing Cantor's paradox.

Quine states Cantor's theorem as "the converse domain of any one-many relation has a subset which does not belong to the domain." Formally,

$$(\forall v)[(\forall y)(\forall z)(\forall w)((z,y) \in v \wedge (w,y) \in v \supset z = w) \supset \\ (\exists x)(\forall y)(y \in x \supset)(\exists z)(z,y) \in v) \wedge (\exists y)(x,y) \in v)]. \tag{3.1}$$

Here, he renders functions in terms of two-place relations, where a relation Rxy is just a set v with the ordered pair (x, y) as a member. We can then read it as "There is a set x, a subset of the converse domain of R, such that for any object y, if it is in x, then it is borne R by something,

[27] Of course, if some other contradiction should be found in the system, then Russell's paradox along with anything else will be derivable as well.

[28] Claims using identity are stratified when the objects on each side of the relation are of the same type in a stratified formula. Quine takes identity as defined in terms of the identity of indiscernibles: $x = y =_{\text{df}} (\forall z)(x \in z \supset y \in z)$. So we could also replace occurrences of identity with this formula, checking the resulting formula for stratification.

[29] W. V. Quine, "On Cantor's Theorem," *The Journal of Symbolic Logic* 2:3 (September 1937), pp. 120–4.

but the set of such *y*'s bears R to nothing, i.e., is not in the domain of R." Hence we have that a set always has more subsets than members.

To prove (3.1), assume that v is a one-many relation, where a one-many relation is a relation such that any two members of the domain of the relation v bearing this relation to the same member in the range must be the same object, that is,

$$(\forall y)(\forall z)(\forall w)((z,y) \in v \wedge (w,y) \in v \supset z = w). \tag{3.2}$$

In other words, such a relation is a function, since for any object in the domain, or input, there is a unique object in the range, or output, to which it bears the relation. By unrestricted comprehension, let x be the set

$$(\forall y)(y \in x \leftrightarrow (\exists z)((z,y) \in v \wedge y \notin z). \tag{3.3}$$

So if $y \in x$, then $y \notin z$, and so $z \neq x$. So from (3.3) and truth-functional logic,

$$(\forall y)(y \in x \supset (\exists z)((z,y) \in v \wedge z \neq x). \tag{3.4}$$

But then by (3.2) and truth-functional logic,

$$(z,y) \in v \wedge z \neq x \supset (x,y) \notin v. \tag{3.5}$$

Therefore by (3.2) and (3.3) and truth-functional logic,

$$(\forall y)(y \in x \supset (x,y) \notin v). \tag{3.6}$$

Now, from (3.3) again,

$$(\forall y)(x,y) \in v \wedge y \notin x \supset y \in x), \tag{3.7}$$

which, by truth-functional logic, is equivalent to

$$(\forall y)(x,y) \in v \supset y \in x). \tag{3.8}$$

so from (3.8)

$$(\forall y)((x,y) \in v \supset (x,y) \notin v), \tag{3.9}$$

which is equivalent to

$$(\exists y)(x,y) \in v. \tag{3.10}$$

From (3.3),

$$(\forall y)(y \in x \supset (\exists z)((z,y) \in v). \tag{3.11}$$

Thus from (3.2) and the unrestricted comprehension principle, (3.1) follows, establishing Cantor's theorem.

As we have seen, unrestricted comprehension led to the set-theoretic paradoxes. Indeed, from this principle, we have seen that we can prove the existence of a universal set, so Cantor's paradox would quickly follow. In Zermelo's system, we can also prove Cantor's theorem, but his restriction on comprehension excludes the existence of a universal set, so there is no reason to worry about the reemergence of the paradox there. But NF, like set theory with unrestricted comprehension, allows for a universal set, so how does it prevent Cantor's paradox?

The proof of Cantor's theorem follows from comprehension, so it will likewise follow from comprehension in NF if the existence condition is stratified. However, it is not stratified. The instance of comprehension used earlier was

$$(\exists x)(\forall y)(y \in x \equiv (\exists z)((z, y) \in v \wedge y \notin z. \tag{3.3}$$

But using the Wiener-Kuratowski definition of ordered pair and checking for stratification, we get

$$(\exists x)(\forall y)(y \in x \equiv (\exists z_n)(\{\{z_n\}_{n+1}, \{z_n, y_n\}_{n+1}\}_{n+2} \in v_{n+3} \wedge y_n \notin z_n), \tag{3.3'}$$

demonstrating that this is an unstratified condition, since y and z must have consecutive subscripts n and $n + 1$. Therefore no such set exists in NF, and Cantor's paradox is blocked. But it also follows now that Cantor's theorem in its stated form is false, and this may appear to weigh heavily against NF as a live option for formalizing set theory, since it now seems to prevent one of Cantor's most important results about set theory. It is central to investigating infinite sets, and losing it would remove much of what made Cantor's theory interesting in the first place.

Quine, however, observes that NF does have an analog to Cantor's theorem. Indeed, it is the same version of the theorem that we can prove for type theory, since there, too, the stated version fails in its violation of the typing restrictions. In type theory, relations can be rendered as sets of ordered pairs in which the members of the pairs must be of the same type. Mathematics, however, frequently requires pairs with members of different types. Without some way around this, mathematics would be all but impossible in type theory. The solution was to use the singleton operation as a way of raising types. So, for example, the pair (x_1, y_2), which is excluded by typing restrictions, can be rendered in type theory in the following way. Again using the Wiener-Kuratowski analysis of the ordered pair, we get $\{\{x_1\}, \{x_1, y_2\}\}$. We then apply the singleton operation to appropriately raise the type so as to get $\{\{\{x_1\}_2\}, \{\{x_1\}_2, y_2\}\}$, which is then the stratified ordered pair $(\{x_1\}_2, y_2)$, a perfectly acceptable statement of type theory.

Similarly, we can now replace the original statement of Cantor's theorem (1) with

$$(\forall v)[(\forall y)(\forall z)(\forall w)((z, \{y\}) \in v \wedge (w, \{y\}) \in v \supset z = w) \\ \supset (\exists x)((\forall y)(y \in x \supset)(\exists z)(z, \{y\}) \in v) \wedge (\exists y)(x, \{y\}) \in v)], \tag{3.1'}$$

which is derivable in NF, since the condition $(\exists z_{n+1})((z_{n+1}, \{y_n\}_{n+1})_{n+2} \in v_{n+3} \wedge y_n \notin z_{n+1})$ is stratified, hence allowing for the existence of the required set. However, where (1) states that the subsets of a set cannot be correlated one-to-one with the set's members, that is, there are more subsets than members, (1′) states that a given set has more subsets than singleton subsets, which has the further consequence that that in general there is no function correlating one-to-one members with their singletons. While this last result may seem especially odd, there are many sets for which such a correlation exists, such as the so-called Cantorian sets, to which we will return in later chapters. To conclude, what Quine shows is that for NF, the original version of Cantor's theorem is false, although an analog of it can be proved in the form that the subsets of a set outnumber the singleton subsets, and that there is no correlation in general between the members of set and their singleton sets.

The next major investigation into the consistency of NF came in 1939 with Rosser's "On the Consistency of Quine's New Foundations for Mathematical Logic."[30] Here, Rosser presents a stronger system, *Q*, and shows that none of the usual methods for producing contradictions are possible in this system. *Q* also yields all of Quine's system NF. Therefore if *Q* is free from contradiction, so is NF.

Rosser begins by remarking that Quine's result that Cantor's paradox is not derivable in NF holds similarly for Rosser's own system *Q*. However, he is able to generalize Quine's findings to show how this result shows the various other set-theoretic paradoxes to be similarly underivable for both *Q* and NF. Rosser observes that the central point of Quine's proof is his showing that there is no function that takes every set to its singleton. That is, there is no set of ordered pairs (x, y) such that x is equal to y where y is the singleton of x, $\{(x, y): x = \{y\}\}$. Indeed, the

[30] J. Barkley Rosser, "On the Consistency of Quine's New Foundations for Mathematical Logic," *The Journal of Symbolic Logic* 4:1 (March 1939), pp. 15–24.

existence of such a set would also yield both the Russell and Burali-Forti paradoxes. For example, consider the following:

$$(\forall R)(\exists \alpha)(\forall x)(x \in \alpha \equiv (\exists y)(x \notin y \wedge \{x\} R y)).$$

Now, instantiating R by $\{(x, y): x = \{y\}\}$ yields

$$(\exists \alpha)(\forall x)(x \in \alpha \equiv (\exists y)(x \notin y \wedge \{x\} = \{y\}),$$

which implies

$$(\exists \alpha)(\forall x)(x \in \alpha \equiv x \notin x),$$

the Russell class. In fact, Rosser concludes that for any unstratified condition φ, we can find a stratified condition ψ with free occurrences of a relation R such that replacing R with $\{(x, y): x = \{y\}\}$ makes ψ equivalent to φ. Since relations are defined as classes of ordered pairs, which in turn are just classes of classes, they depend on stratified comprehension for their existence. Hence attempts to prove the existence of unstratified relations so as to derive the Russell, Cantor, or Burali-Forti paradox fail in NF.[31] So Rosser improves upon Quine's result by showing that this general feature of NF, that is, that in general there is no function taking sets to their singletons, blocks the usual ways of generating the set-theoretic paradoxes in NF, thus furthering the plausibility of NF's consistency.

While a similar argument holds against the Burali-Forti paradox, it is helpful to consider in some detail, as with the Cantor paradox, how exactly it plays out in NF. Unfortunately, it requires a bit more technical machinery than did the discussion of the previous two paradoxes.[32] First, recall that the Burali-Forti paradox is the paradox of the greatest ordinal. Let ON be the set of ordinals that is naturally well ordered by $\leq$, and let Ω be the order-type of $\leq$ restricted to ON. Since Ω is the type of $\leq$ on the ordinals, Ω is greater than all the ordinals in ON, but Ω is also an ordinal, so it must be a member of ON. Hence Ω is both greater than all the ordinals in ON and one of the ordinals in ON.

To begin our discussion of how the Burali-Forti paradox plays out in NF, we must first understand which set-theoretic objects the ordinals are in NF. The cardinals of NF must be Russell–Whitehead cardinals, that is, equivalence classes under equinumerosity, so, for example, the

[31] Ibid., p. 23.

[32] Rosser does this in his later *Logic for Mathematicians* (New York, NY: McGraw-Hill Book Company, 1953), pp. 473–5. Here, I primarily follow Randall Holmes's presentation in *Elementary Set Theory with a Universal Set*, Cahiers du Centre de Logique, vol. 10 (Louvain-la-Neuve: Academia, 1998), pp. 109–10. Holmes is somewhat more accessible than Rosser.

cardinal number of a set x is the set of all objects the same size as x. Similarly, the ordinals of NF must be Russell–Whitehead ordinals. So for a well-ordering (X, R), its ordinal number is the set of all well-orderings order isomorphic to it.[33] Recall that every ordinal α has a successor $\alpha + 1$ such that $\alpha + 1 > \alpha$ and they are naturally well ordered by the relation $\leq$. For α an ordinal, we define an (initial) segment determined by α, written "$seg_{\leq}\{\alpha\}$", as the set $\{\beta \in NO: \beta < \alpha\}$. In a stratified formula, $seg_{\leq}\{\alpha\}$ is one type higher than α. As with the cardinal numbers, type-raising operations will be important if we are to maintain stratification restrictions, so we introduce the operation T. If α is the order-type of (X, R), then $T\{\alpha\}$ is the order-type of $(P_1\{X\}, RUSC(R))$, where $P_1\{X\}$ is the set of singleton subsets of X and $RUSC(R)$ is the set of relational singleton (or unit) subsets of R, that is, the set $\{(\{x\}, \{y\}): xRy\}$. Iterations of the operation T are written "T^n" for n a natural number. A segment $seg_{\leq}\{\alpha\}$, then, is a member of the uniquely determined ordinal $T^2\{\alpha\}$, which is two types higher than α in a stratified formula.

Now we sketch a version of the Burali-Forti paradox. Consider the following inductive argument that for all ordinals α, $T^2\{\alpha\} = \alpha$. Let β be the smallest ordinal such that $T^2\{\beta\} \neq \beta$. For each ordinal $\alpha < \beta$, then, $T^2\{\alpha\} < T^2\{\beta\}$. Furthermore, we see that every ordinal less than $T^2\{\beta\}$ must be $T^2\{\gamma\}$ for some $\gamma < \beta$. But then it follows that $T^2\{\beta\}$ is the smallest ordinal greater than all the ordinals $T^2\{\alpha\}$ for $\alpha < \beta$. That is, $T^2\{\beta\}$ is β itself. Hence it follows that there is no such β such that $\beta \neq T^2\{\beta\}$. So for all ordinals α, $T^2\{\alpha\} = \alpha$. Next, we apply this result to the well-ordering $\leq$ itself. We let Ω be the ordinal containing $\leq$. By this argument, $T^2\{\Omega\} = \Omega$. Since Ω is the ordinal of $\leq$, the ordinal of $\leq$ restricted to $seg_{\leq}\{\Omega\}$ is also Ω. Here arises the contradiction. Since the well-ordering $\leq$ is a strict continuation of $seg_{\leq}\{\Omega\}$, $\leq$ must also include additional ordinals $\Omega + 1$, $\Omega + 2$, and so on greater than the purported greatest ordinal Ω.

Fortunately, Quine's NF does not allow for this argument to go through. Since $T^2\{\alpha\}$ and α are of different types in a stratified formula, $T^2\{\alpha\} = \alpha$ is not stratified and hence does not determine a set. Since we can argue by transfinite induction only with regard to sets, the argument then fails because $T^2\{\alpha\} = \alpha$ and is an unstratified condition. Now we again see Rosser's general point about the singleton function in relation to the set-theoretic paradoxes. There can be no function taking every set

[33] Thomas Forster, *Set Theory with a Universal Set: Exploring an Untyped Universe* (Oxford: Clarendon Press, 1992), p. 44.

to its singleton. If there were, we would then have $T^2\{\alpha\} = T\{\alpha\}$, and $T\{\alpha\} = \alpha$, and thus $T^2\{\alpha\} = \alpha$, allowing the argument to go through, thus yielding the Burali-Forti paradox.

Forster explains the situation slightly differently, observing that with every proper initial segment of ordinals, we can associate two ordinals:

(1) the least ordinal not in X, call it "$L_1(X)$," and
(2) the order-type of X, call it "$L_2(X)$."

While $L_1(X)$ is bigger than every member of X, this is not obviously true of $L_2(X)$. To derive the Burali-Forti paradox, we need to show by induction on the end-extension relation that $(\forall X)(L_1(X) = L_2(X))$. But in NF, the ordinals $L_1(X)$ and $L_2(X))$ belong to different types. Hence $(\forall X)(L_1(X) = L_2(X))$ is unstratified, so the class of counterexamples is not guaranteed to be a set as is required for our induction. In fact, it turns out that there are initial segments of X of the ordinals where $L_1(X) \neq L_2(X)$, so no paradox arises.[34]

As Rosser concludes in his own earlier discussion, "[I]t seems to be the case that there is no danger of deriving a contradiction along any of the known lines until one can handle unstratified relations more effectively."[35] This echoes Quine's point made earlier: With regard to consistency, we are in no better position if we use other versions of set theory. We do not, in general, have absolute consistency results for set theory,[36] so our confidence in its consistency lies in our being unable to derive the paradoxes according to the usual known methods. Consistency for set theory – any set theory – rests upon our not yet having found a means for deriving the set-theoretic paradoxes.

3.3 Some Concluding Remarks on the Significance of Consistency Proofs

The line of thought on the place of consistency proofs for set theory deserves further development. Quine and Rosser's point about NF is not that consistency proofs of any sort are uninteresting or unimportant for foundational studies, and this holds both in the specific case of NF and in

[34] Forster, *Set Theory with a Universal Set*, pp. 44–6.
[35] Rosser, "Consistency", p. 24.
[36] This is not limited to set theory. By Gödel's second incompleteness theorem, no system strong enough for arithmetic or, more specifically, Robinson's Q – Peano arithmetic without the axiom schema of induction – can prove its own consistency. In this sense, consistency proofs for such systems will always be relative to the system in which the consistency proof is carried out, and this system must be at least as strong as the system of which it proves the consistency.

the more general case of mathematics as a whole. Their point is only that since other set theories also lack consistency proofs, NF is no worse off in this respect. To make their claim more precise, let me first consider the sense in which consistency proofs are interesting and important to foundational studies. I will then turn to consider more carefully in this context the sense in which relative consistency proofs for set theory (and this point can be extended to any other branch of mathematics) can serve important mathematical and philosophical aims. Let me first consider some important studies by Georg Kreisel and Michael Resnik on these issues.

In his 1958 "On the Mathematical Significance of Consistency Proofs," Kreisel argues for the importance of consistency proofs for the foundations of mathematics.[37] He sets his discussion in the context of Hilbert's program as outlined in Volume I of Hilbert and Bernays's *Grundlagen der Mathematik*[38] of providing a consistency proof for arithmetic. Kreisel correctly describes this project not as one of traditional foundationalist epistemology, aiming to provide certainty to mathematics through something like self-evident axioms, but as epistemological nevertheless in that the means used for the proof should be "finitistic," or "more 'evident' or more constructive than the principles under discussion." However, he finds two central problems with the project: that they leave unclear exactly what is meant by "constructive" and that the most interesting work found in Hilbert and Bernays's Volume II goes well beyond just a concern with consistency.[39] These defects bring Kreisel to a different program, which he thinks will show more exactly where the mathematical significance of work on consistency problems lies. He resolves the first problem rather quickly by identifying the notion of "constructive" with "recursive."[40] We could of course reject characterizing "constructive" in this way, but I take this as no objection to Kreisel's program. He aims here only to make precise what he found imprecise about Hilbert's original notion; others are free to do likewise.

Having made clear what he will mean by constructive methods, Kreisel then states that the real mathematical gain will be in keeping track of the constructive content contained within mathematical concepts and proofs that are overall non-constructive, particularly within

[37] Georg Kreisel, "On the Mathematical Significance of Consistency Proofs," *The Journal of Symbolic Logic* 23:2 (1958), pp. 155–82.

[38] David Hilbert and Paul Bernays, *Grundlagen der Mathematik*, 2 vols., 2nd edn (New York, NY: Springer-Verlag, 1968/1970).

[39] Kreisel, "On the Mathematical Significance of Consistency Proofs," p. 155.

[40] Ibid.

arithmetic and analysis. The purpose here is not in that constructive consistency proofs will yield greater evidence or reliability to mathematics in contrast to their nonconstructive counterparts, "but in this: they help us to keep track of the constructive (recursive) content of the steps in the (non-constructive) proofs of the system considered. As a result one not only understands 'ordinary' mathematics better but obtains new theorems."[41] Kreisel goes on to provide examples of the sort of new theorems that might be obtained, but the real point of emphasis in the context of Quine's philosophy of set theory is on furthering understanding, as Quine takes this to be a central aim of science generally and so also of philosophy.[42] We will get some indication of this at the conclusion of this chapter and will continue to emphasize this aspect of Quine's philosophy throughout Chapters 4 and 5. But let me first comment on Resnik's more philosophical account of consistency proofs.

Much like Kreisel's article, Resnik's 1974 "On the Philosophical Significance of Consistency Proofs" makes its aim clear in its title.[43] Resnik begins from the standpoint of more traditional foundationalist approaches to the epistemology of mathematics, stating that much of the foundational work in mathematics was initiated in response to the set-theoretic paradoxes. Consistency proofs of the various foundational systems were thought to contribute to this work by justifying these systems as well as the core of classical analysis and set theory. The mathematical contributions of such research, he says, are uncontroversial with Herbrand's work, leading to the completeness proof for quantificational logic being a prime example.[44] However, the philosophical contributions as a response to skeptical worries about mathematics are far less clear. Here, the focus has typically been on absolute consistency proofs, that is, proofs of consistency that do not depend on the consistency of some other system. These latter proofs are the relative consistency proofs, such as those discussed earlier in this chapter with regard to set theory. The basic idea for absolute consistency is to reason directly about the rules of the system itself so as to show that we could never prove from those rules both a statement *p* and its negation *not-p*, although such a proof will still

[41] Ibid., p. 156.

[42] W. V. Quine, *Pursuit of Truth*, rev. edn (Cambridge, MA: Harvard University Press, 1992), p. 2.

[43] Michael D. Resnik, "On the Philosophical Significance of Consistency Proofs," *Journal of Philosophical Logic* 3:1/2 (1974), pp. 133–47.

[44] The relevant papers of Herbrand are his 1930 "Investigations in Proof Theory: The Properties of True Propositions" and his 1931 "On the Consistency of Arithmetic." Gödel, of course, gave the completeness proof itself in his 1930 "The Completeness of the Axioms of the Functional Calculus of Logic." All of these papers are reprinted in van Heijenoort's *From Frege to Gödel*.

depend upon the truth of whatever principles were used for the consistency proof. Resnik says that this is why Hilbert thought that such a proof should rely only on finitistic or constructive means, these being epistemologically less questionable than stronger methods.[45] As is well known, however, Gödel's second incompleteness result shows that such a consistency proof is not available for arithmetic as a whole unless arithmetic itself is inconsistent.[46] What then could be said in response to the skeptic?

Here, Resnik outlines two possibilities. One possibility looks at the consistency of some more elementary theory, such as Peano arithmetic. Again, by Gödel's result, there can be no consistency proof for Peano arithmetic within Peano arithmetic itself. But it could be argued that a consistency proof can be given in some other system that is not stronger than, just different from, Peano arithmetic. Resnik puts forward Gerhard Gentzen's consistency proof for Peano arithmetic as an example of this phenomenon, as have others. If we accept this account of Gentzen's proof, then we may have a response to the skeptics.[47] The other possibility – and I think the philosophically more interesting one – supposes that a system S is a strong system of set theory, say, one that is at least strong enough to reproduce within it all of ordinary mathematics. Then there can be no convincing response to the skeptic by way of a consistency proof. Any such proof would have to take place in a system S' that is at least as strong as the system S, so it would not convince anyone who already questioned the consistency of S itself.[48] What this shows is that the skeptic's demand cannot be met; the skeptic requires too much

[45] Resnik, "On the Philosophical Significance of Consistency Proofs," p. 134.

[46] Resnik raises this question at the outset of "On the Philosophical Significance of Consistency Proofs," saying that it needs further investigation in light of some of Feferman's work (p. 133) and concludes that despite what Feferman says, Gödel's result still shows that a Hilbertian approach to consistency cannot be carried out (p. 145). Feferman forced the reconsideration of the effect of Gödel's result on Hilbert's program by presenting an alternative account of the conditions that a consistency predicate must meet. Resnik concludes that Feferman's alternative diverges too much from our usual understanding of consistency to be a statement of the consistency of arithmetic. The relevant paper here of Feferman's is Solomon Feferman, "Arithmetization of Meta-Mathematics in a General Setting," *Fundamenta Mathematicae* 69 (1960), pp. 35–92.

[47] This account is not universally accepted. The key difference between Peano arithmetic and Gentzen's system is that the latter uses transfinite induction up to the ordinal ε_0. Many critics have argued that at the very least, this does not seem to be in line with the sort of finitism that Hilbert demanded.

[48] Resnik also raises this issue with regard to the consistency of mathematics as a whole. I am not entirely certain what he means here. Such a proof would again require some stronger system within which the consistency proof could be given. But if we are already looking at all of mathematics, then there does not seem to be any further system available. If he means only the usual non-set-theoretic mathematics, then this would be basically the same situation as a set theory strong enough to capture ordinary mathematics.

from mathematics. As Resnik concludes, "We cannot be certain that our axioms are free from contradiction and must treat them as hypotheses which may be abandoned or modified in the face of further mathematical experience. This attitude is taken by many foundational worker who also go on to voice opinions about the *likelihood* [Resnik's italics] that various systems are consistent."[49] As with Kreisel's view of the mathematical significance of consistency proofs, Resnik's assessment of their philosophical significant is in line with Quine's thinking on the matter. In his remark here, Resnik espouses a form of naturalism for mathematics in that it is only within mathematics itself that we can give a consistency proof. The skeptic's demand for something more is incoherent if we hold fast to our scientific standards. What the skeptic asks for is simply impossible.

We have seen already examples of just this sort of attitude in Quine, and I will develop it further in some detail in Chapters 4 and 5. But let me use the topic of consistency proofs here to give some indication of the direction that my account of Quine's general philosophy of set theory will take in the chapters that will follow. Quine is, of course, well aware that there are consistency proofs for set theories. Indeed, by the time of Quine's comment with which I began this section, Gentzen had already proved in 1936 the consistency of the simple theory of types. Some further comment is needed here, however. Russell did not take the axiom of infinity as one of the official axioms of his type theory, in either the simple or the ramified version. Instead, he assumed both this axiom and the axiom of choice only as a hypothesis when needed. It is for this version of type theory that Gentzen proves the consistency, so it is not a system that is strong enough for carrying out foundational work in mathematics.[50]

Still, there are consistency proofs for theories that are strong enough for doing foundational work. These cases bring us back to relative

[49] Resnik, "On the Philosophical Significance of Consistency Proofs," p. 145.

[50] Gentzen's proof can be found in his "The Consistency of the Simple Theory of Types" reprinted in *The Collected Papers of Gerhard Gentzen*, ed. M. E. Szabo (Amsterdam: North-Holland, 1969), pp. 214–22. I do not know how readily available in the United States Gentzen's work was in the 1930s. Quine was, on the whole, well informed about mathematical research in Europe during this time, thanks to his postdoctoral travel grant. He used it to go to Europe to meet Rudolf Carnap and to learn firsthand the most up-to-date research in the foundations of mathematics. Quine was also an early proponent of natural deductions systems, which were due to Gentzen as well. However, Quine explicitly stated that he learned of such systems not from Gentzen's work but from meeting Stanislaw Jaskowski in Warsaw in 1933 (W. V. Quine, *The Time of My Life*, Cambridge, MA: MIT Press, 1985, p. 194). Quine himself sketches such a consistency proof for simple type theory in his *Set Theory and Its Logic*, rev. edn (Cambridge, MA: Harvard University Press, 1969), pp. 263–4.

consistency proofs, in which we prove the consistency of one system relative to some other system that is at least as strong as the theory for which we are proving consistency. As we just saw in the work of Resnik, with such proofs, our confidence in the proof of consistency is only as great as our confidence in the consistency of the system within which we carried out the proof. This is again Quine's point that there are no absolute consistency results for set theory. So if we proved Peano arithmetic to be consistent relative to some system of set theory, say, ZFC, we would have proved it to be consistent on the assumption that ZFC is itself consistent.[51] Speaking of rather strong mathematical theories, such as the various set theories we have been considering in these first three chapters, Quine himself later characterized the situation as follows:

> Consistency becomes more questionable, and consistency proofs become more urgent. But they also become more difficult and less conclusive. The more elaborate the apparatus to which a consistency proof appeals, the more room there is for questioning the consistency of that apparatus itself and so for questioning the conclusion that the apparatus was used to establish. In the more speculative axiomatizations of set theory the most we can usually aspire to in the way of a consistency proof is a proof that one such system is consistent if another somewhat less mistrusted one is consistent.[52]

If we think that a consistency proof should answer to more traditional sorts of foundational concerns over the epistemology of mathematics – say, for example, with how we can ever know that mathematical knowledge is certain – these relative consistency results will be of little use, as Quine notes.

Much of the point of the account of set theory that I have offered so far has been to warn against these sorts of foundationalist aspirations. I have been presenting a historical and philosophical account of set theory that shows it developing from within mathematics itself and not as serving as a sort of first philosophy for mathematics. This view also leaves room for seeing the actual importance of relative consistency results. They are important if, as Kreisel urged, we seek understanding. Indeed,

[51] Again, there are other methods for showing Peano arithmetic to be consistent. Gentzen proved Peano arithmetic to be consistent by using transfinite induction up to the ordinal ε_0. These methods must also in some sense go beyond what is available in Peano arithmetic itself. Otherwise, we would violate Gödel's second incompleteness theorem; see his "On Formally Undecidable Propositions of *Principia Mathematica* and Related Systems I," in *Collected Works*. For Gentzen's proof, see his "The Consistency of Elementary Number Theory," in his *Collected Papers*, pp. 132–213.

[52] Quine, *Set Theory and Its Logic*, p. 286.

Quine takes this to be not just a mathematical aim but a central aim for philosophy generally.[53] Relative consistency results contribute to it in various ways, as Kreisel noted. Quine describes others. For one, they help to bring out conceptual connections of a sort between different mathematical theories in that consistency proofs can show us the consistency strength of one theory over another. In addition, this allows us to track the strength of the various mathematical theories we might be willing to assume. Quine highlights these features in his 1963 *Set Theory and Its Logic*, observing that consistency proofs give us a precise proof-theoretic measure of strength.[54] As I have already remarked upon, by Gödel's second incompleteness theorem, a reasonably strong set theory cannot prove its own consistency unless that set theory is itself inconsistent. But, as I remarked earlier in the chapter, when we can prove the consistency of one set theory in another set theory, then the latter theory is the stronger.[55] Quine goes on to explain that a method then readily presents itself for generating a stronger set theory from any given set theory. We can add to a set theory a new axiom by way of Gödel's coding of formulas into statements of arithmetic – what is known as Gödel numbering – that states the consistency of the previous theory. We might then even form a single system of this progression of axioms, but then the original method might be continued, again yielding stronger and stronger set theories. Quine concludes by observing that this is the pattern of transfinite recursion: "The strength of a system of accumulated axioms specified along these lines is limited only by the potentialities of transfinite recursion – the ordinal strength, in turn – of the medium of communication in which we are specifying that infinite fund of axioms."[56] Throughout the remaining chapters, we will return to the drawing out of conceptual connections, such as this one, and keeping track of assumptions as a central aspect of Quine's philosophy of set theory and of his philosophy more generally.

We have also, as Quine alludes to, that a relative consistency proof can help to further our confidence in a theory that has not been as well studied as some other theory. This characterizes the current situation

[53] Although Quine speaks of the aim of science as understanding, there is no firm boundary between science and philosophy in his view, so this accurately characterizes much of his philosophical undertakings as well. On this point, see his *Pursuit of Truth*, p. 2.

[54] Quine notes in his discussion that there are other measures of strength, perhaps related, such as ordinal strength; see his *Set Theory and Its Logic*, sec. 45.

[55] Quine, *Set Theory and Its Logic*, pp. 324–5.

[56] Ibid., p. 325. He includes a footnote with further references to this method of generating stronger and stronger set theories.

with regard to his NF.[57] In comparison to Zermelo's set theory and its variants, NF has received far less scrutiny by logicians and mathematicians. If a consistency proof could be found for NF within a variant of Zermelo's set theory, researchers in the field would likely feel much more comfortable with NF as part of mainstream set theory. Again, we see here connections between theories emerging and furthering our understanding of set theory generally. All of this, as we will continue to see in the remaining chapters, is fully in line with a Quinean approach to the philosophy of set theory. So now, with Quine's NF accounted for and some suggestion of the philosophy behind it, let us turn to considering and developing his philosophy of set theory more fully. In particular, we will see in Chapter 4 that Quine's views on the philosophy of set theory are a microcosm of his philosophy as a whole.

[57] I should stress that this was not the situation when Quine first presented NF. In the mid-1930s, set-theoretic research was a much more pluralistic undertaking than it is now. Only sometime around the 1960s did Zermelo's set theory emerge as dominant and become nearly synonymous with set theory. This topic will be discussed further in Chapters 6 and 7.

PART II

Quine, Set Theory, and Philosophy

CHAPTER 4

Quine's Philosophy of Set Theory

In the previous chapters, I presented the development of set theory through W. V. Quine's New Foundations (NF). While I did not entirely avoid philosophical discussion, I turn more directly in this chapter and the next to philosophy – specifically to Quine's philosophy of set theory. In Chapters 1–3, I emphasized to varying degrees what I have identified as a conception of set theory as explication.[1] In this chapter, I will develop this approach to the philosophy of set theory in some detail, as found in Quine's early work. This will continue in Chapter 5, where I will focus on his mature philosophy. In earlier chapters, we saw the origins of this approach to set theory in the work of set theory's founders. While we find definite aspects of it in the work of Georg Cantor, Bertrand Russell, and Ernst Zermelo, Quine was really the first to fully develop this conception as an explicit philosophical approach to set theory.[2] He sums up the view nicely in his 1941 "Whitehead and the Rise of Modern Logic." Reflecting on the various attempts to come to terms with the set-theoretic paradoxes, Quine observes,

> But a striking circumstance is that none of these proposals, type theory included, has an intuitive foundation. None has the backing of common sense. Common sense is bankrupt, for it wound up in contradiction. Deprived of his tradition, the logician has had to resort to mythmaking.

[1] I have contrasted this approach to set theory with that of conceptual analysis, which strives to discover a single correct conception of set theory that can somehow be drawn out of the very concept of set itself. Most commonly, this approach is seen in attempts to derive the axioms of set theory from the iterative conception of set. I will return to this topic in Chapters 6 and 7. In *Set Theory and Its Philosophy* (New York, NY: Oxford University Press, 2004), Michael Potter draws a distinction between regressive and intuitive approaches to set theory (p. 34). This seems to correspond pretty closely to what I mean by the distinction between explication and conceptual analysis, respectively.

[2] As I mentioned in the Introduction, Quine attributes the notion of explication to Rudolf Carnap. I take it that he and Carnap would largely agree on approaching set theory as a process of explication, but I think Quine was the one to fully develop this as a philosophical account of set theory. In many ways, this is the project that inspired him to begin philosophy, and it remained a significant part of his philosophical work throughout his career.

> That myth will be best that engenders a form of logic most convenient for mathematics and the sciences; and perhaps it will become the common sense of another generation.[3]

Here, we get a clear and succinct statement of the key elements of Quine's philosophy of set theory: that set theory after the paradoxes will be largely a conventional affair guided by pragmatic considerations of convenience as to which set theory will best serve as a framework for science and mathematics.[4] Though less vividly, Quine adhered to this view already in his 1934 *A System of Logistic*, remarking that type theory serves its purpose if it blocks the paradoxes, and it is a view that was certainly implicit in his 1932 dissertation.[5] In both of these works, Quine does not see the theory of types as capturing something like the essence of sethood; for Quine, there is no essence to be had. As is indicative of the tradition of set theory as explication, he aimed merely to restrict set theory enough to prevent contradiction while leaving enough strength in place for it still to serve as a framework for the rest of mathematics and so also for "the abstract structure of all science."[6] More important, this is an attitude that became a constant in Quine's philosophy more generally. It is the idea that science is a theory that we construct, though as such, it is no less real. This attitude is what he aims to convey in the oft-quoted remark that "[t]o call a posit a posit is not to patronize it."[7] This idea informs my reading of Quine throughout. I will return to it in Chapter 5, where I discuss the philosophical significance of his approach to set theory and the foundations of mathematics as aiming to clarify and simplify the mathematical portion of our conceptual scheme. Such

[3] W. V. Quine, "Whitehead and the Rise of Modern Logic," in *Selected Logic Papers*, p. 27.

[4] In the quote, Quine speaks of the best logic. During this period, he still spoke of set theory as part of logic, but only out of piety for the usage of his predecessors Gottlob Frege, Russell, and Carnap; see Quine's "Reply to Hao Wang" in *The Philosophy of W.V. Quine*, eds. Lewis Hahn and Paul Arthur Schilpp, expanded edn (Chicago, IL: Open Court, 1986), pp. 644–8. Quine's later insistence on distinguishing set theory from first-order quantification theory does not, I think, indicate any significant change in his philosophical views. However, pursuing this topic further here would, take me too far from the subject matter at hand.

[5] W. V. Quine, *A System of Logistic* (Cambridge, MA: Harvard University Press, 1934), p. 19. His dissertation was eventually published in its original form with a preface added by Quine as *The Logic of Sequences: A Generalization of Principia Mathematica* (New York, NY: Garland Publishing, 1990).

[6] W. V. Quine, *The Time of My Life: An Autobiography* (Cambridge, MA: MIT Press, 1985), p. 85

[7] Quine, *Word and Object*, p. 22. I should note that the idea of myth making also appears throughout Quine's philosophical work and is not limited to the mathematical sciences. It is, for Quine, an aspect of constructing scientific theories generally. See, for example, his "On What There Is," in *From a Logical Point of View: Nine Logico-Philosophical Essays*, 2nd rev. edn (Cambridge, MA: Harvard University Press, 1980), pp. 18–19; "Two Dogmas of Empiricism," in *From a Logical Point of View*, p. 45; and "Posits and Reality," in *The Ways of Paradox and Other Essays*, rev. and enlarged edn (Cambridge, MA: Harvard University Press, 1976), p. 251.

clarification and simplification are much of what Quine takes as the aim of philosophy as a whole.

The structure of this chapter is as follows. I begin in Section 4.1 by returning to Russell; he was, as we will see, Quine's first important philosophical influence, particularly with regard to his work in logic and the foundations of mathematics. It is in starting from Russell, I will argue, that Quine develops his approach to set theory as explication. Previously, I have mostly emphasized Russell's adherence to the view of set theory as explication and its more pragmatic aspects. In Section 4.1, I will discuss other features of his philosophy that yield some tension in his view. I will argue in Section 4.2 that Quine's starting point in Russell is an attempt to come to terms with these tensions by rejecting the extra-mathematical aspects of Russell's philosophy so as to arrive at a philosophy of set theory that answers only to concerns from within mathematics itself. In short, we have here Quine's naturalism restricted to the philosophy of set theory.[8] I will focus in Section 4.2 on Quine's dissertation, "The Logic of Sequences," and its published version, *A System of Logistic*. This work develops directly out of Quine's attempts to clarify and simplify Russell's logicism, and it is here that we see Quine working out the tensions of Russell's foundational work in mathematics. We also start to get a better sense of the unique philosophical vision motivating Quine's work: the view that philosophy is largely concerned with the clarification and simplification of our conceptual scheme, a topic that will receive full treatment in Chapter 5. Finally, in Section 4.3 I will return to Quine's work in set theory up through to the period surrounding the publication of "New Foundations for Mathematical Logic," but in contrast to the discussion in Chapter 3, the emphasis here is much more on the philosophical aims that this technical work in set theory is meant to achieve.[9] Much of the discussion here concerns a frequently overlooked paper from 1938, "On the Theory of Types."[10] It is here that Quine's rejection of more metaphysical approaches to set theory emerges explicitly.

[8] I do not emphasize it as much here, but I also think that this influence carried over to shape Quine's philosophical outlook as a whole. For more on this, see my "Quine, Russell, and Naturalism: From a Logical Point of View," *Journal of the History of Philosophy* 53:1 (2015), pp. 133–55. Some of what follows overlaps with the content of that paper.

[9] W. V. Quine, "New Foundations for Mathematical Logic" with supplementary remarks. In *From a Logical Point of View*, pp. 80–101.

[10] W. V. Quine, "On the Theory of Types," *Journal of Symbolic Logic* 3:4 (1938), pp. 125–39. I have also discussed this paper in my "The Significance of Quine's New Foundations for the Philosophy of Set Theory," *The Monist* 100:2 (2017), pp. 167–79.

4.1 Background in Russell

In Chapter 2, I stressed one strand of Russell's thought in approaching the foundations of mathematics: that rooted in the tradition of set theory as explication. In this section, I will emphasize a different strand: that rooted in more traditional philosophical concerns within which Russell's views on the foundations of mathematic answer more to concerns outside of mathematics itself. With this backdrop in place, we will be able to see more clearly the ways in which Quine moves against Russell by urging that set theory be developed in response to concerns coming from within mathematics and the sciences more generally.

Russell famously opens *Principles of Mathematics* by declaring his twofold aim.[11] First, he intends to show "the proof that all pure mathematics deal exclusively with concepts definable in terms of a very small number of fundamental logical concepts, and that all its propositions are deducible from a very small number of fundamental logical principles." His second aim is "the explanation of the fundamental concepts which mathematics accepts as indefinable," the purpose being "to see clearly … the entities concerned, in order that the mind may have that kind of acquaintance with them which it has with redness or the taste of a pineapple." He describes this second aim as "purely philosophical."[12] The purpose of such a reduction of mathematics to logic – what has become known as "logicism" – was not somehow to better assure us of the certainty of mathematics; rather, it was part of a complex argument against Idealism. Russell wanted to show that, contrary to the claims of the Idealists, at least some of our nonmetaphysical knowledge does not ultimately lead to contradiction; he thought he could achieve this by way of reducing mathematics to logic.[13] We see here one of the most striking and important features of Russell's philosophy: that his technical mathematical work is intimately entwined with his philosophical aims.[14] It is a theme to which we will return throughout this chapter to show how Quine's own philosophy emerges from this entanglement. We will see Quine, in a sense, liberating logicism from many of the philosophical

[11] Bertrand Russell, *Principles of Mathematics*, 2nd edn (London: George Allen and Unwin Ltd., 1937 [first edition 1903]), p. xv.

[12] Ibid.

[13] For full details on Russell's emergence from and reaction to Idealism, see Peter Hylton's comprehensive study *Russell, Idealism, and the Emergence of Analytic Philosophy* (Oxford: Clarendon Press, 1990) and his "Logic in Russell's Logicism," in his *Propositions, Functions, and Analysis*, pp. 49–82.

[14] Hylton, *Russell, Idealism, and the Emergence of Analytic Philosophy* emphasizes this aspect throughout his account of Russell's philosophy.

burdens Russell placed on it, though Quine, too, will have philosophical aims upon which he brings technical methods to bear.

Because the British Idealists took metaphysics to be the only subject of which there could be knowledge that is noncontradictory, all science short of metaphysics was considered to be ultimately contradictory. Much of their work attempted to make explicit the contradictions found in the mathematical and natural sciences. Russell himself began in this tradition but rejected it sometime in the late 1890s. His 1903 *Principles of Mathematics* emerged out of work that began during his Idealist period as part of an attempt to rework Hegel's encyclopedia by bringing it into line with the sciences.[15] Russell explains in his preface to *Principles* that his project had its origins in trying to cope with some difficulties in the foundations of dynamics, which led him to consider the philosophy of continuity and the infinite and so also the foundations of mathematics. Because mathematics is the basis of all science, if Russell could show that mathematics was noncontradictory, he would have shown that the Idealists were wrong to contend that metaphysics constituted the only noncontradictory science. The reduction to logic would show this. In Russell's hands, then, logicism was part of a broader project of providing a consistent account of science generally. More specifically, it showed, in contrast to Idealism, that a consistent account of absolute space was ready to hand. Truth, then, did not come in degrees as the Idealists claimed but rather could be shown to be absolute. Whether in physics, mathematics, or metaphysics, truth was everywhere the same. Again, Russell thought that this could be successfully carried out by showing that mathematics ultimately reduces to logic.[16] As we will see, though, this depends crucially on the reduction being to something that could plausibly be considered to be logic.

In this context, Russell's logic was not just a technical development that he then brought to bear on Idealism. Rather, logic itself presupposed a particular metaphysics, a metaphysics that Russell drew directly from G. E. Moore:

> On fundamental questions of philosophy, my position, in all its chief features, is derived from Mr G. E. Moore. I have accepted from him the non-existential nature of propositions (except as happen to assert

[15] For more on Russell's Idealism and his rejection of it, see Hylton, *Russell, Idealism, and the Emergence of Analytic Philosophy*; see also Nicholas Griffin, *Russell's Idealist Apprenticeship* (Oxford: Clarendon Press, 1991).

[16] For further detail on this point, see Hylton, *Russell, Idealism, and the Emergence of Analytic Philosophy*, pp. 179–80.

> existence) and their independence of any knowing mind; also the pluralism which regards the world, both that of existents and that of entities, as composed of an infinite number of mutually independent entities, with relations which are ultimate, and not reducible to adjectives of their terms or of the whole which these compose. Before learning these views from him, I found myself completely unable to construct any philosophy of arithmetic, whereas their acceptance brought about an immediate liberation from a large number of difficulties which I believe to be otherwise insuperable. The doctrines just mentioned are, in my opinion, quite indispensable to any even tolerably satisfactory philosophy of mathematics, as I hope the following pages will show.[17]

This atomistic metaphysics committed to abstract objects, that is, propositions, combined with the view that truth and falsity are absolute, Hylton labels "Platonic atomism."[18] According to Platonic atomism, our knowledge of the world is direct and unmediated, a view that runs directly contrary to that of the Idealists, who claimed that all of our knowledge is conditioned by various conceptual structures.[19] These structures, they claimed, are objective, corresponding to the way the world actually is. But this is difficult to defend. If all knowledge is mediated, then the claim itself that the world does in fact correspond to these conceptual structures must also be mediated by other conceptual structures. But then the question returns: What can we say of the objectivity of these other conceptual structures? And this question will continue to arise for further such conceptual structures, resulting in an infinite regress. While the Idealists did in fact have responses to such charges, neither Moore nor Russell saw them as sufficient to sustain the view that such mediated knowledge could be objective knowledge. We could not, on the basis of the Idealist view, have genuine true knowledge of how the world really is.[20] Against this, Russell, following Moore, denies this central claim of Idealism and instead takes our knowledge of the world to be direct and unmediated.[21] Our perception of the world is of how the world really is. Perception here should be understood in a

[17] Russell, *Principles of Mathematics*, p. xviii.

[18] My account of Russell's metaphysics in the period of his logicism follows that of Hylton. See his *Russell, Idealism, and the Emergence of Analytic Philosophy* as well as his *Propositions, Functions, and Analysis*. Many of the essays in *Propositions, Functions, and Analysis* deal with relevant topics, but here I follow the brief account presented in "The Theory of Descriptions," pp. 189–94.

[19] As Hylton notes, in Kant's case this account brushes over the distinction that he draws between intuitions and concepts; see Hylton, *Propositions, Functions, and Analysis*, p. 190, fn. 13.

[20] Hylton, *Propositions, Functions, and Analysis*, pp. 190–1.

[21] The views that Russell takes from Moore are put forward perhaps most clearly in Moore's "The Nature of Judgment," in *G. E. Moore: The Early Essays*, ed. Tom Regan (Philadelphia, PA: Temple University Press, 1986), pp. 59–80.

very broad sense, since both Russell and Moore thought that we could have something like perception of ordinary everyday physical objects as well as perception of abstract objects. Indeed, we have already seen this in how Russell explains the very aim of his logicist reduction: "*to see* clearly ... the entities concerned, in order that the mind may have that kind of acquaintance with them which it has with redness or the taste of a pineapple" [my emphasis]. Acquaintance is the relation that gives the mind direct unmediated access to what is outside of it.

So among the ideas that Russell takes from Moore is that of direct unmediated access to the world, and it is in this context that propositions enter as central to Russell's philosophy. Returning to the previously quoted passage, in line with the objectivity gained by this direct access to the world, Russell then states that propositions are entities that are independent of any knowing mind. As he understands them, propositions are nonlinguistic, nonmental abstract entities and, roughly speaking, the content expressed by a declarative sentence. In making a judgment, we have acquaintance with propositions, which have among their properties being true or false; in a true judgment, we have acquaintance with a true proposition, and in a false one, we have acquaintance with a false proposition. So again, in contrast to the Idealists, truth and falsity do not depend on any sort of mediation by ideas in a mind but rather result from a direct relationship with a proposition that has the requisite property of truth or falsity.

Given that Russell's view of propositions is atomistic, it is in this sense that he regards the world as a plurality made up of "an infinite number of mutually independent entities." Propositions themselves are complex and so can be analyzed into their basic constituents, which Russell calls "terms." Hylton explains that Russell views this process of analysis as analogous to something like chemical decomposition. For example, the proposition "Socrates is wise" can be decomposed into its two basic components: Socrates and wisdom, since, in paradigm cases, the proposition actually contains the objects that it is about.[22] Russell concludes the quoted passage by saying that he believed that only by adopting this metaphysics could he successfully develop his philosophy of arithmetic. It is at this point, I think, that a certain tension begins to emerge in Russell's philosophy with regard to his notion of analysis, and this will be crucial to the understanding of Quine I will present. The tension emerges between Russell's philosophical commitments and his more

[22] Hylton, *Propositions, Functions, and Analysis*, pp. 9–10, 191–2.

technical mathematical approach to carrying out his logicism. On the one hand, he takes mathematical results to be relevant to philosophical problems and even to solve them. On the other hand, he still holds that there is some special further task for philosophy to do beyond mathematics, namely, that it acquaints us with the indefinables of mathematics. It is here that we find the tension in his notion of analysis.

Russell describes pure mathematics in *Principles of Mathematics* as that class of propositions of the form "*p* implies *q*," where *p* and *q* are propositions each containing at least one variable, the same in each proposition, and having no constants but logical constants. The logical constants are notions that are definable in terms of implication, the notion of *such that*, the relation of a term to the class of which it is a member, the notion of a relation, and other further notions involved in propositions of the form just described. Russell includes the notion of truth among these, though it is not itself a constituent of propositions. He then aims to show that this definition suffices to account for what has traditionally been thought of as pure mathematics. Furthermore, this definition is

> not an arbitrary decision to use a common word in an uncommon signification, but rather a precise analysis of the ideas which, more or less unconsciously, are implied in the ordinary employment of the term. Our method will therefore be one of analysis, and our problem may be called philosophical – in the sense, that is to say, that we seek to pass from the complex to the simple, from the demonstrable to its undemonstrable premisses.[23]

So the notion of analysis is crucial, and it is this method that will allow us to become acquainted with the indefinables of pure mathematics "as the necessary residue in a process of analysis."[24] In addition, analysis aims to make precise common notions that had not previously been formulated with the clarity and rigor demanded by science. What, then, is this process of analysis?

On first glance, it seems that analysis is merely the technical reduction of mathematics to logical notions. As Russell explains, it will be through "the labours of the mathematicians themselves" that he will be able to obtain certainty and clarity with regard to his questions about the nature of number, infinity, space, time, and motion and about mathematical inference itself. In reducing such questions to questions of pure logic, we

[23] Russell, *Principles of Mathematics*, p. 3.
[24] Ibid., p. xv.

find exact knowledge about mathematics. However, this is not the whole story. Russell continues, explaining that the philosophy of mathematics had previously been just as controversial and unprogressive as other branches of philosophy. Philosophy demanded a meaning for mathematics, but mathematics had no answer. Now, he claims, mathematics does have an answer "so far at least as to reduce the whole of its propositions to certain fundamental notions of logic." Again, it seems that the reduction of mathematics to logic answers what used to be the concerns of philosophy. For example, it is now fairly common to think that in its technical reduction, logicism has shown us what the natural numbers really are. But then Russell adds, "At this point, the discussion must be resumed by Philosophy."[25] Beyond the technical reduction of mathematics to logic, Russell believes that there is still some further task for philosophy to carry out. This becomes all the more apparent if we consider his contrasting of two notions of definition – the philosophical and the mathematical – which reflect respectively the two aspects of his notion of analysis described earlier.

Russell first introduces this distinction between philosophical and mathematical definition in discussing Giuseppe Peano's logical work:

> It is necessary to realize that definition, in mathematics, does not mean, as in philosophy, an analysis of the ideas to be defined into constituent ideas. This notion, in any case, is only applicable to concepts, whereas in mathematics it is possible to define terms which are not concepts. Thus also many notions are defined by symbolic logic which are not capable of philosophical definition, since they are simple and unanalyzable.[26]

He goes on to explain that in mathematics, we define by simply picking out some fixed relation to a fixed term, which only one term can have. The basic idea is that mathematical definition consists in giving necessary and sufficient conditions for an entity to fall under a concept.[27] Russell then distinguishes philosophical definitions, explaining, "The point in which this differs from philosophical definition may be elucidated by the remark that the mathematical definition does not point out the term in question, and that only what may be called philosophical insight reveals which it is among all the terms there are."[28] So while necessary and sufficient conditions are enough for mathematical definition, which brings

[25] Ibid., pp. 4–5.
[26] Ibid., p. 27.
[27] Ibid., pp. 26–7, 111; Hylton, *Russell, Idealism, and the Emergence of Analytic Philosophy*, pp. 187–8.
[28] Russell, *Principles of Mathematics*, p. 27.

us back to the idea of an analysis as merely rendering an imprecise concept precise, philosophical definition requires something more. It seems that the mathematical definition can pick out any particular object from among several satisfying the requisite conditions, but philosophical definition goes deeper, telling us in some sense which objects the defined term *really* is or what its *true* meaning is.[29] Russell has this sense of definition in mind when he talks of analysis as the method for carrying out his philosophical aim of leading us to acquaintance with the objects of pure mathematics.[30]

Such an understanding of philosophical and mathematical definitions fits well with Russell's later remarks in *Principles of Mathematics* on defining cardinal numbers. He defines the cardinal number of a class as the class of all classes similar to the given class, where two classes are similar if their members can be put into one-to-one correspondence. What all similar classes have in common, then, is their cardinal number.[31] So a number is just a class of a particular sort. Although this definition of cardinal number succeeds, mathematically speaking, Russell admits some philosophical worries about the connection between classes and predicates. In a previous section, Russell had explained that a philosophical definition is "the analysis of the idea ... into constituent ideas" but that this notion "is only applicable to concepts," unlike definition in mathematics.[32] Here again, we have the idea of analysis as a decomposition into simple constituents, which is, according to Russell, what a philosophical definition should give us. Furthermore, it is this notion of definition that leads us to acquaintance with the real reality, a metaphysical reality, whereas mathematical definition does not necessarily do this.

But then a philosophical difficulty remains in that Russell does not know whether the appropriate concepts can even be found to identify with the numbers, that is, whether, through a process of analysis, he has truly arrived at the indefinables of mathematics. It may be that mathematical definition does not *necessarily* lead us to the sort of acquaintance that philosophical definition does but that it may in some cases. How, then, do we identify such cases? Russell himself appears at least in some

[29] Russell makes this point in discussing the analysis of betweenness in section 196 of *Principles of Mathematics*.

[30] Hylton makes this point in *Russell, Idealism, and the Emergence of Analytic Philosophy*, p. 232.

[31] Russell, *Principles of Mathematics*, pp. 113–15.

[32] Russell, *Principles of Mathematics*, p. 27. Russell also repeats this notion of philosophical definition at the outset of his discussion of cardinal numbers, remarking that "philosophically, the word *definition* has not, as a rule, been employed in this sense [that is, as merely satisfying necessary and sufficient conditions]; it has, in fact, been restricted to the analysis of an idea into its constituents" (p. 111).

places to believe that this concern does not really matter, at one point remarking of the philosophical notion,

> This usage is inconvenient and, I think, useless; moreover it seems to overlook the fact that wholes are not, as a rule, determinate when their constituents are given, but are themselves new entities …, defined, in the mathematical sense, by certain relations to their constituents. I shall therefore, in future, ignore the philosophical sense, and speak only of mathematical definability.[33]

This is at first puzzling if philosophical definition is supposed to yield the kind of acquaintance with the simple constituents of mathematics that Russell demands of his project. His earlier remarks on philosophical definition were in no way disparaging and seemed to carve out an important role for philosophy. I think that what Russell intends to say here is that philosophical definitions are useless when it comes to discussing the technical, mathematical details of his reduction. He still thinks that there is a role for philosophical definition, but it is in reflecting on what the mathematical definition has accomplished that it comes into play. For example, have we achieved acquaintance with the actual indefinables of mathematics? But increasingly, this yields also a tension between the philosophical and the mathematical. Once the mathematical definition has presented the requisite sorts of objects for mathematics, what further analysis is there to do? What further question is there to answer? As we will see, Russell continues to look at both notions of definition, but increasingly, he takes the mathematical notion to do the work of the philosophical. Yet he remains committed to the philosophical, metaphysical foundation of his project. It is unclear what is left for the philosophical notion of definition to do. Russell's account of definition is not the only place where the tension between the mathematical and the philosophical arises. The tension between these two aspects continues into his later adoption of the axiom of reducibility.[34]

In Chapter 2, we saw evidence of Russell's approach to set theory as explication in his resolving of the paradoxes. But recall that as a consequence of the hierarchy of types, Russell was forced to adopt the axiom of reducibility to recover the reduction of mathematics to logic. Let me

[33] Russell, *Principles of Mathematic*, pp. 111–12.

[34] Another such problem arising out of Russell's distinction between the philosophical and the mathematical is that of the unity of the proposition. For more on this, see my "Quine, Russell, and Naturalism." From Quine's perspective, there is also a fundamental lack of clarity in the intensional basis for Russell's logicism, which again lies on the philosophical rather than the mathematical side of his work. I will discuss this further later in the chapter.

develop this point a bit further here. In resolving the paradoxes, Russell motivated the distinctions of order and the resulting types by way of his vicious circle principle as applied to propositional functions, but he ultimately justified his system by its success in doing away with the paradoxes. This solution is both mathematical and pragmatic – mathematical in that it yields technical resolution to the paradoxes and pragmatic in that the criterion of success for his system is that the paradoxes are blocked in a way that still allows him to develop his reduction of mathematics to logic. I should emphasize that the solution is tied up in the philosophical aspects of his project in that it is rooted in the metaphysics of propositions, which he brought to bear on his Idealist opponents. In this vein, it is notable that Russell does not give up Platonic atomism as a philosophical position despite initially coming upon his famous paradox in terms of propositional functions. Furthermore, we should observe that, in contrast to F. P. Ramsey and Quine (who, as we saw in Chapter 2, both distinguished the semantic from the class paradoxes), Russell does not opt for the simple theory of types, blocking just the set-theoretic paradoxes, that is, blocking just those paradoxes that are relevant to mathematics. He vastly complicates his universe with the system of orders and types rather than giving up his fundamental metaphysics. Because logic presupposed the metaphysics of propositions, a solution to the paradoxes means eliminating the semantic and set-theoretic paradoxes all at once.

As we saw previously, once Russell introduces these type distinctions, he still requires some way to talk about all properties if his logic is going to be sufficiently strong to yield mathematics. To this end, he puts forward the axiom of reducibility. Recall that the axiom states that for every propositional function, there is an extensionally equivalent predicative propositional function, that is, an extensionally equivalent propositional function of the lowest order compatible with its arguments.[35] While Russell clearly takes the existence of propositional functions as part of his logic, reducibility is problematic in its existence claims. Logically speaking, there seems to be no reason to assume the existence of the predicative propositional functions that the axiom puts forth. It seems a fact about the world that we could as easily reject as accept. This raises a difficulty about the nature of logic for Russell. Indeed, one of Wittgenstein's most persistent criticisms of Russell's logic was questioning the justification for

[35] Alfred North Whitehead and Bertrand Russell, *Principia Mathematica*, vol. 1, 2nd edn (Cambridge: Cambridge University Press, 1925), p. 56.

reducibility.[36] The absolute truth of logic functions as a crucial component of Russell's attack on Idealism, but it is hard to argue on purely logical grounds that this axiom is unconditionally true. Furthermore, the axiom does not seem to fit with any traditional characterization of logic as self-evident, a priori, or analytic. Even taking a Quinean stance and saying that the truths of logic are just among the truths that we hold to most firmly in our web of belief would seem to do little for the logical status of reducibility. A crucial axiom in Russell's logicism then fails in its logicality.

Russell recognizes this and in fact does not try to ground the axiom in pure logic. As we have seen, in the 1908 paper in which he first presented his theory of types, he merely appeals to the axiom's success in allowing for the deduction of certain key statements of arithmetic to proceed, such as mathematical induction.[37] In *Principia Mathematica*, he develops this approach further, taking it to apply generally in the case of adopting any axiom:

> The reason for accepting an axiom, as for accepting any other proposition, is *always* largely inductive, namely that many propositions which are nearly indubitable can be deduced from it, and that no equally plausible way is known by which these propositions could be true if the axiom were false, and nothing which is probably false can be deduced from it. If the axiom is apparently self-evident, that only means, practically, that it is nearly indubitable; for things have been thought to be self-evident and have yet turned out to be false [my emphasis].[38]

Here, Russell's justification for reducibility is of the wholly pragmatic sort, relying on how successful the system in which it is used captures mathematics. In his preface to *Principia Mathematica*, he makes the same point, reminiscent of Zermelo in his 1908 paper[39] (see Chapter 2), stating that "the chief reason in favor of *any* theory on the principles of mathematics must always be inductive, i.e. it must lie in the fact that the theory in question enables us to deduce ordinary mathematics" [my emphasis].[40] Russell develops this approach to the foundations of mathematics most

[36] For example, see Ludwig Wittgenstein, *Tractatus Logico-Philosophicus* (New York, NY: Routledge, 1990 [1922]), pp. 6.1232–6.1233; see also his *Cambridge Letters: Correspondence with Russell, Keynes, Moore, Ramsey, and Sraffa*, eds. Brian McGuinnes and Georg Henrik von Wright (Malden, MA: Blackwell Publishers, 1997), pp. 35, 53–4, 59–61.

[37] Russell, "Theory of Types," pp. 167–8.

[38] Whitehead and Russell, *Principia Mathematica*, p. 59.

[39] Zermelo, Ernst, "Investigations into the Foundations of Set Theory I," in *From Frege to Gödel: A Source Book in Mathematical Logic*, ed. Jean van Heijenoort, 1879.

[40] Ibid., p. v. Russell expresses a very similar sentiment in the earlier *Principles of Mathematics*, p. xviii.

fully in a 1907 lecture, "The Regressive Method of Discovering the Premises of Mathematics."[41] Here, he sets out to explain in what sense comparatively difficult and obscure propositions can be the premises for propositions, which are, on the whole, comparatively obvious, as is typically the case with the foundations of mathematics. Within this context, he "emphasize[s] the close analogy between the methods of pure mathematics and the methods of the sciences of observation."[42] For example, he says that it seems absurd to prove a claim such as "2 + 2 = 4" from the rather complex and unintuitive axioms of a logical system, such as the one put forward by Gottlob Frege in his 1893/1903 *Grudgesetze der Arithmetik*[43] or the one that Russell himself was in the process of working out at the time of this lecture. Instead, as he would later urge in *Principia Mathematica*, Russell argues that we believe the premises because they yield consequences that we already believe to be true. This, he says, is at the heart of the inductive method and so is really no different from finding general laws in the natural sciences. The starting point in both cases is "a body of propositions of which we are fairly sure." We then work backward to the premises for these claims, which, though in some sense are more complicated, are in another sense logically simpler.[44]

Among the advantages of such an arrangement are that it allows us to isolate the pervading element of falsehood found in any theory and that it lends organization to our knowledge generally.[45] For example, in coming to terms with Russell's paradox, Russell says that the regressive method leads us to look for the least obvious among Frege's logical premises. In doing so, we are led from making drastic revisions to Frege's system, for example, denying the law of contradiction, to instead looking for some less obvious premises to revise, such as his basic law V.[46] Still, obviousness itself is no substitute for any kind of Cartesian indubitability.

[41] Bertrand Russell, "The Regressive Method of Discovering the Premises of Mathematics," in *Essays in Analysis*, Lackey, pp. 272–83. The lecture was not published until 1972, so Quine could not have read it as he was developing his early views. It is striking how closely Quine's own views end up coinciding with Russell's as put forth in this lecture. Quine, in this sense, can be understood as following out to their full conclusion the remarks that Russell did publish in *Principia Mathematica*, which Quine read as an undergraduate.

[42] Russell, "Regressive Method," p. 272.

[43] Gottlob Frege, *Basic Laws of Arithmetic*, trans. Philip A. Ebert and Marcus Rossberg (New York, NY: Oxford University Press, 2013).

[44] Ibid., pp. 273–4. For a full account of Russell's views on analysis, see Paul Hager, "Russell's Method of Analysis," in *The Cambridge Companion to Bertrand Russell*, ed. Nicholas Griffin (New York, NY: Cambridge University Press, 2003), pp. 310–31.

[45] Russell, "Regressive Method," p. 275.

[46] Ibid., p. 278.

It always remains a matter of degree where we simply prefer to maintain the more obvious propositions over the less obvious ones, and even those propositions with the highest degree of obviousness might turn out false. Furthermore, when we find that one obvious proposition is deducible from another, both become more certain than they were in isolation. For this reason, Russell thinks that a complicated deductive system with many obvious parts gains a total probability of near certainty. "Thus," he concludes, "although intrinsic obviousness is the basis of every science, it is never, in a fairly advanced science, the whole of our reason for believing any one proposition of science."[47]

In *Principia Mathematica*, this talk of obviousness is replaced by the notion of self-evidence, as is seen in the quote with which we began this discussion of an inductive justification. Again, though, Russell's talk of self-evidence does not carry the burden of certainty that is usually attributed to it in more foundationalist approaches to epistemology. In this sense, it differs little from walking out into the rain and declaring that it is raining. Furthermore, Russell also observes that self-evidence does not yield any strong justification for an axiom, since some apparently self-evident truths have turned out to be false, specifically Frege's basic law V, the source of Russell's paradox. Regardless of self-evidence, which reducibility surely does not have, Russell qualifies the inductive reason for accepting an axiom, that is, that it yields the desired results and no false ones, as the *chief* reason for accepting it. On these grounds, reducibility is perfectly acceptable as part of Russell's logical system. But the tension in Russell's views on the philosophical and mathematical aims of his project become apparent here. In the service of an argument against Idealism, logicism needs to be a reduction to something that we can all agree to label as logic. In particular, logic must be absolutely true. Reducibility makes this highly problematic, since it seems just as likely to be true as to be false. As a merely formal technical reduction of mathematics to logic, however, Russell's system succeeds, thus making available the pragmatic justification that he takes to justify any axiom, including reducibility. As we will see, it is this aspect of Russell that Quine latches onto in developing a philosophy of set theory as explication. Noticeably absent in this approach is any more traditional concerns about the nature of logic.

I will argue later in this chapter that Quine resolves this tension between the philosophical and the mathematical by fully adopting

[47] Ibid., p. 279.

Russell's view that technical results can solve philosophical problems. In doing so, though, he will let the further task that Russell claims for philosophy fall by the wayside. It is in fully committing himself to the idea that mathematics can solve philosophical questions that, I claim, we have the origins of Quine's naturalistic philosophy.[48] In short, we have Quine taking the best methods of the science of his day – in this case, those of the new mathematical logic – and using them to address philosophical concerns. Let me turn now directly to Quine.

4.2 Early Developments: From "The Logic of Sequences" to New Foundations

Quine's 1932 dissertation, "The Logic of Sequences," was his first major work in set theory. In it, he reworked roughly the first 500 pages of Whitehead and Russell's *Principia Mathematica*, generalizing their account of propositional functions and relations. He begins by explaining that he views the dissertation as allied with *Principia Mathematica* but that it is more comprehensive in its generalization of *Principia Mathematica's* account of propositional functions and relations. *Principia Mathematica* has a system of monadic propositional functions to do the work of classes along with a separate theory of dyadic relations so that parallel axioms and theorems must then be given for each of the two distinct realms. Furthermore, Russell's system did not generalize to *n*-adic relations; it is impossible to prove theorems in general for *n*-adic relations without first specifying the value of *n*. Since dyadic relations are enough for reducing mathematics to the system of *Principia Mathematica*, this is as far as Russell goes. By instead employing a system of sequences, Quine's system allows for classes and relations to be treated singly rather than as two independent realms. As a result, Quine does not need to reprove parallel theorems for relations once he has done so for classes. Furthermore, such sequences may be of any arbitrary length, so theorems for relations are proved generally for any *n*-adicy. In addition to this, Quine observes also that his system is more economical in primitive ideas, though also more elaborate in its postulates. Quine also believes that he has achieved greater elegance in his definitions. As he would later observe, all of this went toward the philosophical project, which "aspired,

[48] I will discuss this to some extent in the rest of this chapter and Chapter 5, but for an account that is more focused on this idea, see my "Quine, Russell, and Naturalism."

like *Principia*, to comprehend the foundations of logic and mathematics and hence the abstract structure of all science."[49]

As we have seen, the system itself is a variation on Russell's theory of types, though with some important differences. As with Russell in his 1908 paper, Quine offers no justification for adopting types beyond the theory's success at resolving the paradoxes while still providing enough strength for the reduction of ordinary mathematics to logic.[50] In line with the approach to set theory that Quine lays out in the quote with which we began this chapter, throughout his dissertation he presents no justification for his reworking of *Principia Mathematica* beyond the elegance and simplicity that he achieves. As I have already noted, one of Quine's central improvements over the original system of *Principia Mathematica* is his generalization of it so that theorems can be proved for relations of any *n*-adicy. More significant philosophical departures from Russell are Quine's own take on the axiom of reducibility and his extensionalizing of Russell's hierarchy of propositional functions.

In his dissertation, Quine, too, adopts the axiom of reducibility but not as a formal postulate of his system. Instead, he says of this axiom that "[w]e are regarding it as an informal assumption justificatory of a certain notational convention."[51] Rather than taking it as a formal postulate, Quine recommends instead adopting it only as a prosystematic convention that says that all functions that would be nonpredicative in *Principia Mathematica* be reinterpreted in his system as predicative functions. Still, the assumption of reducibility is needed because otherwise, his convention of reinterpretation might yield notation that expresses impossible notions, that is, his system would symbolize predicative functions that have no entities corresponding to them. Quine claims that his approach makes better sense of the axiom than does Russell's. In *Principia Mathematica*, the axiom does receive full symbolic expression, but Quine explains that the status of predicative functions remains ambiguous, since the notion is never formally defined or taken as an

[49] Quine, *Time of My Life*, p. 85. We will see more of how this view develops in Quine's philosophy in discussing his *Word and Object* and *Set Theory and Its Logic*. For now, we should just note that it was part of Quine's philosophical project from the very start.

[50] In the dissertation, Quine actually offers less justification than this. He does not remark on the paradoxes at all but rather adopts the theory of types without comment and just shows the theory's adequacy for capturing ordinary mathematics; see *The Logic of Sequences*, preface, chapters XII and XIII. In the published version of the dissertation, the 1934 *A System of Logistic*, Quine does remark on type theory as having the purpose of resolving the paradoxes in addition to showing the theory adequate to deriving ordinary mathematics; see p. 19.

[51] Quine, *The Logic of Sequences*, pp. 5–6.

explicit primitive of the system. By rendering all functions as predicative and relegating any talk of order to prosystematic discussions of his system, Quine eliminates any ambiguity in the status of the notion of a predicative function and the associated notation "$\varphi!x$."[52]

In the published version of his dissertation, the 1934 *A System of Logistic*, reducibility receives much less discussion. Quine perhaps recognized more clearly that because his system is fully extensional, all that matters is extensional equivalence among his functions.[53] The distinction between predicative and nonpredicative functions, then, has no role to play in his system. Two functions with the same extension are, from the standpoint of his system, the same function. There is no need for any talk of differences in order.[54] This brings us to the second significant divergence from Russell's system: that in both Quine's dissertation and the published version of it, he never strays from a completely extensional logic. From the outset, he makes clear that his talk of propositional functions is interchangeable with talk of classes and relations.[55] While Quine makes little of this move in his dissertation, it is a significant break with Russell, who saw his intensional propositional functions as lying at the basis of the extensional classes upon which he would then construct the rest of mathematics.

That this is an important philosophical move against Russell cannot be emphasized enough. Russell's adoption of propositional functions as the basis for his logic was not without thought or motivation.[56] First, these were the only sorts of objects that Russell took to have the requisite generality to be properly logical. Previously, he had granted this status also to classes, but with the discovery of the class paradox, he recognized that he would have to subject classes to some sort of restriction. Second, since classes could be defined contextually in terms of propositional functions, unresolved questions about their nature – whether they were intensional or extensional and how a class of many objects could be treated as a single object – no longer required answers. Third, as we saw in Chapter 2, Russell saw the ontological distinctions between types as flowing from, and hence justified by, the distinctions between propositional functions and individuals, which emerged from his account of

[52] Ibid., pp. 7–8.

[53] Quine would also soon publish separately his views on the axiom of reducibility; see his "On the Axiom of Reducibility," *Mind* 45:180 (1936), pp. 488–500.

[54] Quine, *A System of Logistic*, p. 186.

[55] Ibid., pp. 18–19, 31–2.

[56] I follow Hylton in my brief account here. For more detail on Russell's reasons for taking propositional functions as fundamental and the difficulties arising from doing so, see Hylton, *Russell, Idealism, and the Emergence of Analytic Philosophy*, pp. 288–327.

propositions. Hence the resolution of the paradoxes and the view of the universe as structured into a hierarchy of types rested upon taking propositional functions as primary. Russell had no similar justification for the hierarchy if he instead took classes as basic.

Quine, of course, rejects this intensional basis for logicism and instead treats propositional functions extensionally, identifying them with the classes of objects they are true of: their extensions.[57] The move comes without immediate justification, but it is implicit in the entire aim of the dissertation, showing that Quine's system is adequate for yielding the entire system of *Principia Mathematica*.[58] Hence there is no need to adopt intensional propositional functions; their extensional counterparts are enough. These motivations become all the more apparent in the published version of the dissertation, *A System of Logistic*, as Quine explains,

> This assimilation of propositional functions to classes and of predication to membership represents no actual impoverishment of logistic, but only the elimination of useless lumber: for, as will subsequently be shown, all theorems of PM involving function variables or predication can be proved in the present system under the indicated manner of translation.[59]

But why should we prefer the extensional to the intensional? Quine says very little of the distinction in the dissertation, but again he elaborates his motivations in *A System of Logistic* in discussing his extensional treatment of propositions. He warns that an intensional account may force us "to leave the terra firma of algorithmic logic and tread more metaphysical ground."[60] In particular, he complains that propositions are rather difficult to pin down as identical. An extensional account, however, provides a clear identity criterion. We can easily identify propositional functions as the same when they have the same extensions.[61]

This all might seem rather ad hoc in comparison to what Russell thought a well-motivated development of the logic required for carrying out logicism. He thought just this, reflecting later in life on Quine's systems of logic being such that not "even the cleverest logician would have

[57] Quine, *The Logic of Sequences*, pp. 3–5.

[58] Ibid., chapter 13.

[59] Quine, *A System of Logistic*, p. 31. In *Our Knowledge of the External World* (New York, NY: W. W. Norton, 1929), p. 51, Russell talks of using logical techniques to clear away "metaphysical lumber."

[60] Quine, *A System of Logistic*, p. 33.

[61] For a later and somewhat autobiographical account of Quine's preference for extensionality, see his "Confessions of a Confirmed Extensionalist," in *Future Pasts: The Analytic Tradition in Twentieth-Century Philosophy*, eds. Juliet Floyd and Sanford Shieh (New York, NY: Oxford University Press, 2001), pp. 215–21.

thought of [them] if he had not known of the contradictions."[62] But we have also seen Russell himself thinking in just this way about his axiom system, justifying it inductively on grounds that it "enables us to deduce ordinary mathematics."[63] It is this strand of Russell that Quine latches onto. In adopting an extensional view of logic, Quine pushes aside difficult, if not hopeless, worries that are external to mathematics and thus opens the way to a philosophy of mathematics that takes place entirely within mathematics itself. For deducing ordinary mathematics, only the extensions matter; all else is "useless lumber." In short, we can see Quine beginning to work out a philosophy of mathematics in which there is no first philosophy. He elaborates on this issue further when he comes to consider explicitly "the more philosophical side of logic."[64] Having introduced propositions as sequences on purely technical grounds, he now considers how this account plays out with regard to a number of philosophical issues, chief among them Russell's intensional account of propositional functions.

Given that Russell thought only an intensional account of propositional functions could ensure the logicality of his system, Quine's extensional account is no small departure. From it, we get a very clear indication of what Quine takes philosophy to be. He gives no argument from the nature of set theory or logic more generally for this move but emphasizes only the clarity that is achieved and the fending off of more metaphysical flights. Observing that materially equivalent propositions – those that are both true or both false – are indistinguishable within the propositional calculus, Quine explains that Frege simply accepted that there are only two propositions: the true and the false. The result is no doubt strange, but Quine approves of the motivations lying behind it:

> If on the other hand Frege's conclusions be rejected, what is propositional identity to involve over and above material equivalence? In seeking more exacting conditions of identity than material equivalence, danger is to be apprehended of having to leave the *terra firma* of algorithmic logic and tread more metaphysical ground; the usual formal procedure, therefore, is to avoid the question by handling material equivalence as such and never mentioning the identity of propositions [Quine's italics].[65]

[62] Russell, *My Philosophical Development*, p. 61.

[63] Whitehead and Russell, *Principia Mathematica*, p. v. Russell remarks similarly in *Principles of Mathematics* (see p. xviii). More specifically, as we saw earlier in this chapter and in Chapter 2, he attempts to ground the axiom of reducibility in such a way.

[64] Quine, *A System of Logistic*, p. 32.

[65] Ibid., p. 33.

However, Quine's account in terms of sequences improves upon this further by distinguishing more finely between propositions while still remaining fully extensional:

> It is a singularity of the doctrine of propositions as sequences that it give a definite meaning to the identity of propositions which goes far beyond the trivial condition of material equivalence, and that in so doing the doctrine continues to confine itself to strictly logistical territory, and, indeed, to import nothing exceptionable to the most ardent extensionalist.[66]

Identity of propositions is merely identity between sequences, where two sequences x,y and z,w are identical just in case $x = z$ and $y = w$. Here, we get a much more refined account of propositional identity than that of just material equivalence yet one that still differentiates among propositions that appear to convey the same meaning. For example, taking α as the class of all cities now smaller than Chicago, β as the class of all cities now larger than Paris, y as Paris, and w as Chicago, Quine explains that the propositions α,y and β,w are materially equivalent and seem to convey the same information but are not the same proposition. He concludes, "The doctrine of propositions as sequences thus not only endows propositional identity with a thoroughly definite and satisfactorily discriminative meaning, but does so quite without commerce beyond the strictly extensional medium of mathematical logic."[67] Here we have what Quine sees as the real philosophical import of his account: the clarity it brings to logic by keeping it fully extensional. There is no recourse to some mysterious metaphysical realm of, for example, meanings. There is just the clear mathematical criterion for identity among sequences. Indeed, in Quine's own recollections, this "bid for clarity" was one of the dissertation's central contributions toward the philosophical project of comprehending the abstract structure of all science. For this reason, he saw it as a considerable improvement over Russell's intensional basis for logic.[68]

Some philosophers might complain that Quine's moves here ignore the important philosophical issues concerning the nature of propositions. But Quine would deny any deeper nature to propositions than what his

[66] Ibid.

[67] Ibid., p. 34.

[68] Quine, *Time of My Life*, p. 85. For many, this might not seem a philosophical achievement worth achieving, but for Quine, it is. The other achievements that Quine cites (strict adherence to use and mention and economy of definition and notation) might seem similarly underwhelming for many philosophers. For Quine, they are central to his philosophical endeavor, and this is very telling about his aims in the philosophy of set theory and in philosophy generally.

account already provides for. Furthermore, such criticism would fail to recognize how Quine sees technical methods as coming to bear on philosophical problems. His approach might appear to be just a technical trick applied to an otherwise deep philosophical problem.[69] But for Quine, such technical maneuvers are at the heart of approaching philosophical problems, deep or otherwise. In line with what would become the naturalism of his mature philosophy, Quine's account of propositions here appeals to the best that current science has to offer in order to make sense of the initially hazy or even mysterious notion of a proposition. Expressing his basic agreement with Whitehead over the nonassertiveness of propositions in contrast to judgments, Quine writes, "The doctrine of propositions as sequences stands in striking agreement with Whitehead's point of view; it presents a definite technical entity fulfilling just the demands which he makes of a proposition."[70] Here, we have in mind certain conditions that propositions should satisfy. Within the context of a logical system such as that of *Principia Mathematica* or of *A System of Logistic*, propositions must, for example, be bearers of truth and falsity, be manipulable in accord with the rules of the logical system, and so on. As long as the entities that are proposed as propositions can do this, there is no reason not to take them to be propositions. There is no deeper essence that such entities need to capture. It is precisely this approach that Quine would later identify as paradigmatic of philosophical analysis.

To bring us back to our general theme, we see Quine in his early work in logic and set theory further developing what I have been calling set theory as explication. His logical systems do not develop from what he sees as the nature of the essence of sets but rather answer to concerns from within the mathematical sciences. In this way, we can see his early logical work as a precursor to his mature naturalism applied to philosophy more generally.

[69] In response to a criticism of this sort from Kripke about Quine's criterion for ontological reduction, he writes, in "Responses," in *Theories and Things* (Cambridge, MA: Harvard University Press, 1981), p. 175:

> One of Kripke's moral precepts deplores "the tendency to propose technical criteria with the aim of excluding approaches that one dislikes" ... He notes in illustration that I adopted a criterion of ontological reduction for no other reason than that it "includes well-known cases and excludes undesired cases ..." I protest that mine was expressly a quest for an objective criterion agreeing with our intuitive sorting of cases. This is a proper and characteristically philosophical sort of quest, so long as one knows and says what one is doing.

[70] Quine, *A System of Logistic*, p. 33.

4.3 New Foundations as Philosophy of Set Theory

From the time of his dissertation up to the 1937 "New Foundations for Mathematical Logic," Quine published roughly fifteen additional papers. Most of these focus on the technical development of logic, although there are two overtly philosophical papers among them: "Ontological Remarks on the Propositional Calculus" and "Truth by Convention."[71] Still, much like the dissertation and *A System of Logistic*, these technical papers reflect the heart of Quine's philosophy of set theory. We see in all of these papers a continual striving to clarify and simplify Quine's basic logical system, what he would himself later describe as an attempt "to settle on a sanest comprehensive system of logic."[72] For example, in his 1936 "Concepts of Negative Degree," he observes, "As with imaginary numbers, no intuitive meaning is contemplated for these concepts of negative degree; their utility is structural simplification."[73] In the 1937 "On Derivability," he provides an account of derivability without appealing to classes: "[I]t is the purpose of this paper to renounce all use of class variables, relation variables, or function variables, and find an equivalent construction in terms of weaker devices."[74] As part of his clarificatory work, Quine always looks to what a theory must assume and what it can get by without, as he does in this paper. While these concerns might be understood as typical of any mathematical undertaking, for Quine they are the beginnings of what he would come to see the aim of philosophy: a striving to clarify our conceptual scheme as a whole.[75]

Throughout this period, Quine continues to focus his attention on the theory of types, frequently citing what he sees as the two main drawbacks of Russell's system: its arbitrary grammatical exclusions of otherwise seemingly intelligible combinations of signs and the reduplication of

[71] Quine, "Ontological Remarks on the Propositional Calculus." Reprinted in W. V. Quine, *The Ways of Paradox and Other Essays*, rev. and enlarged edn (Cambridge, MA: Harvard University Press, 1976), pp. 265–72; "Truth by Convention." Reprinted in *The Ways of Paradox and Other Essays*, pp. 77–106.

[72] W. V. Quine, "Autobiography of W. V. Quine," in *The Philosophy of W. V. Quine*, eds. Lewis Edwin Hahn and Paul Arthur Schilpp (La Salle, IL: Open Court), p. 17.

[73] W. V. Quine, "Concepts of Negative Degree," *Proceedings of the National Academy of Sciences of the United States of America* 22:1 (1936), p. 42. Quine describes concepts of negative degree by way of concepts with positive degree. The basic idea is to introduce the idea of *n*-adic relations, where *n* can be less than zero. Quine take concepts to be the general term for predicates (one-place relations) and relations of any *n*-adicy.

[74] W. V. Quine, "On Derivability," *Journal of Symbolic Logic* 2:3 (1937), p. 114.

[75] For more the beginnings of Quine's philosophy from his early logical work, see my "Quine, Russell, and Naturalism."

objects at each level of the type hierarchy.[76] Many of the papers in this period try to come to terms with these drawbacks. Both of the 1936 articles "A Theory of Classes Presupposing No Canons of Type" and "Set-Theoretic Foundations for Logic" provided ways of accounting for the type hierarchy and its associated grammatical restrictions in terms somehow more primitive than by way of assuming a universe stratified into types.[77] Quine's hope was that he could then define types contextually:

> Only such locutions issue from these [contextual] definitions as would be countenanced by the familiar theory of types. The theory of types itself, as an *index expurgatorius*, therefore becomes superfluous; [fn. deleted] the forms which it brands meaningless are already meaningless here in the strict sense of not being defined in terms of the primitives [Quine's italics].[78]

The locutions that are banned as ungrammatical are banned simply as a result of what could be defined contextually in terms of the primitives of the particular system that is adopted. As we saw in Chapter 2, Quine's concerns here with type theory continued into his 1937 "New Foundations for Mathematical Logic." Indeed, it was again these oddities about types that drove Quine to develop the system presented there. While Chapter 2 focused mostly on the technical details of the system and Quine's own early development of it,[79] let me now situate Quine's aims into their more philosophical setting.

As we saw, NF emerged out of Quine's reflections on the systems of both Russell and Zermelo, preserving something like the type restrictions of the one theory along with the unrestricted variables of the other. In doing this, Quine adopts as a decidedly philosophical account of set theory, elements of which we saw in the work of Cantor, Russell, and Zermelo, specifically, the idea that set theory serves to explicate the notion of set rather than, through a process of conceptual analysis, identifying the uniquely correct notion of a set. Central to Quine's approach, and to this approach in general, is to develop a set theory that takes set theory's mathematical contributions as the criterion of success for the particular system set forth. Let me here set out the general philosophical concerns that

76 Quine makes these points repeatedly in discussing his dissatisfaction with types; see, for example, his "The Inception of 'New Foundations,'" in *Selected Logic Papers*, pp. 286–7.

77 Quine, "A Theory of Classes Presupposing No Canons of Type," *Proceedings of the National Academy of Sciences* 22 (1936), pp. 320–6.

78 Ibid.

79 Specifically, in Chapter 2, we focused on Quine's account of how Cantor's theorem can be reconciled with a universal set.

Quine sees NF as developing out of and answering to. In the chapters that follow, I will respond in more detail to more strictly mathematical concerns, such as the development of arithmetic and the axiom of choice.

Quine presents very little in the way of explicit philosophical motivations in "New Foundations for Mathematical Logic" itself. Certainly, there is the implicit concern with simplifying our theory of sets, which then further allows us to simplify our scientific theory as a whole so as to further our understanding of the abstract structure of science. "New Foundations for Mathematical Logic" demonstrates this not by pursuing some preordained notion of a set, but by developing a logic that Quine finds "most convenient for mathematics and the sciences." As in the papers leading up to "New Foundations for Mathematical Logic," Quine highlights the oddities of Russell's type theory: the reduplication of objects, including the null and universal sets along with the numbers themselves; the lack of absolute complements and reduplication of the class algebra for each type; and, overall, the difficulties of restoring severed connections that result from this reduplication. As I explained in Chapter 3, Russell's practice was not to adhere to explicit type restrictions but rather to leave formulas typically ambiguous, with the understanding that appropriate type indexes could always be made explicit if need be. To arrive at the system of NF, Quine follows such ambiguity through to its natural conclusion: If the theorems and proofs of them can be stated without making the types explicit, then there is little reason to think of the levels that come with the theory of types as inherent to the set-theoretic universe. Instead, Quine takes the typing restrictions as a syntactic test, which he applies only to the set existence axiom schema. Here, as we have seen, any claim of set existence must pass the test of stratification; that is, the membership condition on the right-hand side of the biconditional must obey the restrictions of the theory of types. This technique eliminates the various oddities that Quine disliked in Russell's original theory. But there are still deeper philosophical motivations here that emerge more clearly in an interesting and overlooked paper from 1938, "On the Theory of Types." Here, Quine makes clear the philosophical implications of NF for the philosophy of set theory generally.

Again, Quine explains how a system such as his NF does away with the oddities of reduplication found in Russell's type theory, driving the point home especially well by considering the status of the empty set:

> This reduplication is particularly strange in the case of the null class. One feels that classes should differ only with respect to their members, and this is obviously not true of the various null classes. A unique null class indeed

> still seems permissible, vacuously, if we think only of the requirements that members be alike in type. However, other requirements of type theory would be violated. For example, we want the null class to be included in each class; hence, inasmuch as it is regarded as meaningless to relate classes of unlike types by inclusion, we need a new null class to be included in each class of new type.[80]

Quine also adds here that types have difficulties proving the theorem of arithmetic that, for all finite natural numbers n, $n \neq n + 1$. Quine explains that we might proceed by starting with the empty set and the universal set of any specific type, and then, by way of standard set-theoretic operations, we can produce the class consisting of none, one, the other, or both of these sets as elements, giving us a set of four members. This procedure can then be repeated until, for any given n, we get a set of at least n members, which the proof of the theorem will require. But finding such sets will require us to ascend into higher and higher types, so the theorem might actually be false if we do not ascend high enough. Whitehead and Russell were, of course, aware of this difficulty and so assumed an axiom of infinity for each type to overcome it.[81] But of this solution, Quine says, "Some such axiom is in any case presumably needed for the theory of infinite numbers; but that it should be needed for proving finite inequalities is an anomalous effect merely of the theory of types."[82]

However, Quine makes clear in this article that there is a deeper philosophical point to his reworking of Russell's theory of types, a point that is left implicit in nearly all of his other writings where he reflects on the move from the theory of types to NF. Quine frames his discussion in this paper by observing that the theory of types consists of two aspects, which we should be careful to distinguish – an ontological, or metaphysical, doctrine and a formal restriction – and it is the first of these that Quine sees as the source of the unnatural effects of type theory.[83] He explains this ontological, or metaphysical, aspect as

> stipulat[ing] that if an individual is a member of a class x, then x must be composed exclusively of individuals; if an individual is a member of a member of a class x, then x must be composed exclusively of classes composed exclusively of individuals; and so on. Individuals are said to be of *type* 0, and classes of objects of type n are said to be of *type* $n + 1$; and in

[80] Quine, "On the Theory of Types," p. 131.
[81] Whitehead and Russell, *Principia Mathematica*, vol. 2, pp. 203–4.
[82] Quine, "On the Theory of Types," p. 132.
[83] Ibid., pp. 125, 129.

> these terms the theory of types amounts, in its ontological aspect, to demanding that all the members of a class be alike with respect to type.[84]

I take it as significant that Quine describes this aspect as metaphysical, since his use of this term during this period tends to follow Rudolf Carnap's with all of its negative connotations.[85] Given the peculiarities that result from the adoption of types along with the use of typical ambiguity in actual practice, Quine sees something extramathematical and so, more generally, extrascientific in the commitment to types as an ontological doctrine, asserting that the set-theoretic universe actually *is* stratified into the levels of the type hierarchy. His point becomes all the more clear in comparing what he has to say about the second aspect of type theory, that is, its formal restriction.

According to this formal aspect, a formula is meaningless rather than false if it includes any subformula in which the membership relation is flanked by two variables that are not of consecutive ascending types or if the subset or identity relation is flanked by variables that are not of the same type. The meaningful atomic formulas then take the following forms, respectively, $x_n \in x_{n+1}$, $x_n \subset y_n$, or $x_n = y_n$.[86] While often expressed by reference to the type ontology, the formal aspect does not dictate that the set-theoretic universe be in fact stratified into such a hierarchy. This emerges clearly when Quine considers Russell's actual practice of leaving the formulas typically ambiguous. As we have seen already, Russell dispenses with any type indices, as "is more usual and more convenient," leaving the variables "typically ambiguous." The idea then is that indices could be added to the variables to show that the formulas actually do adhere to the type restrictions, but there need be no talk of the set-theoretic universe actually being stratified into levels.[87] Quine explains by describing the actual proof that Whitehead and

[84] Ibid., p. 129.

[85] I would say that Quine's latter uses of this term are most often also in the critical spirit of Carnap, although the story becomes more complicated after his explicit break with Carnap over analyticity and what Quine describes as the blurring of the boundaries between speculative metaphysics and natural science. Quine's use of the term "doctrine" may also be significant, in that he uses this years later in "Carnap and Logical Truth" in discussing the linguistic doctrine of logical truth, another metaphysical aspect to be purged from scientific philosophy. Quine, "Carnap and Logical Truth," Reprinted in W. V. Quine, *The Ways of Paradox and Other Essays*, rev. and enlarged edn (Cambridge, MA: Harvard University Press, 1976), pp. 107–32.

[86] For ease of exposition, I have stated the formal aspect rather informally here, as is standard in the literature and as Quine himself initially does ("On the Theory of Types," p. 129). He presents a completely formal statement of it as (IV) on pp. 130–1 of "On the Theory of Types."

[87] Quine, "On the Theory of Types," p. 130.

Russell give of the arithmetical theorem that for all natural numbers n, $n \neq n + 1$. By leaving their variables typically ambiguous, Whitehead and Russell can prove this theorem in full generality without restricting it to any particular level of the type hierarchy and without having to add an axiom of infinity. "Such a proof, though admitted by the apparent formalism of *Principia Mathematica* and related systems," Quine remarks,

> seems to involve an abuse of typical ambiguity: a theorem is unconditionally asserted which, judged merely on its internal structure, admits determinations of type not covered by the proof. Hence Whitehead and Russell did not choose the easy way; indeed, to avoid being deceived into this fallacious sort of argument they even brought in a heuristic notation of suffixes for keeping track of the range of types covered by a proof [fn. omitted]. No such precautions were explicit in the initial formalism of their system, and indeed it would be a matter of some complexity to incorporate them explicitly. Obviously the abuse of typical ambiguity would be much more convenient. Further, despite its apparent lack of cogency, this practice seems never to yield any intrinsically undesirable theorems.[88]

Whitehead and Russell's proof shows that such "undesirable theorems" are blocked not by the ontological aspect of types but by its formal aspect. Since their proof avoids all talk of specific types to which the variables are assigned, it does not force a commitment to any claim that the objects of the theory actually exist only at specific levels of the hierarchy. The formal aspect, however, still plays a key role by preventing the kind of self-membership that would yield Russell's paradox. So Quine concludes,

> The awkward situations thus far considered actually depend, not on the formal aspect of type theory, but only on the ontological aspect. Let us then try abandoning the ontological aspect altogether, retaining only the formal restrictions: for if the theory of types is adequate at all as a safeguard against contradictions, it must be adequate in its formal aspect alone.[89]

He then adopts this formal restriction for the specifying conditions found in set abstracts or, for NF, in the comprehension axiom schema.

With the abandonment of a typed ontology, classes can now contain members of mixed type, for example, a class may contain as members both individuals and classes, and even some instances of self-membership

[88] Ibid., p. 132.
[89] Ibid.

are allowed, such as the fairly intuitive cases $V \in V$ and $\Lambda \notin \Lambda$, both of which type theory's ontological aspect prevented. We also gain absolute complements rather than just complements within a specific type. In addition, constants such as "V," "Λ," and the numerals cease to be typically ambiguous. Indeed, typical ambiguity is dispensed with altogether, since there are no longer any types to be ambiguous about. This allows Whitehead and Russell's proof of the theorem that for all natural numbers n, $n \neq n + 1$ to go through without raising any awkward questions about the abuse of types. Only the formal aspect of types is retained with no suggestion of any sort that the universe is actually typed into levels.[90] As Quine concludes, "The type ontology was at best only a graphic representation or metaphysical rationalization of the formal restrictions; and though some such rationalization may well be desired, it seems clear in particular that the type ontology afforded less help than hindrance."[91] The type restrictions remain merely as a syntactic test for allowable formulas but do not support any stronger ontological, or metaphysical, claim about the structure of the set-theoretic universe.

Of course, present-day set theory also takes the form of cumulative types. We have seen this already in discussing Zermelo's set theory and its related systems. But as with type theory, Quine has observed oddities here too – for example, that Zermelo's theory also rules out a universal set as well as absolute complements, though not in the same way that type theory does.[92] Something very much like the ontological doctrine that Quine described earlier continues to hold sway in contemporary philosophy of set theory, living on in what has become known as the iterative conception of set, as embodied in the axioms of Zermelo's theory. But much like the situation with types, Quine would see the levels of the iterative hierarchy, though cumulative, as having their origins in a metaphysical view about sets rather than in the mathematical theory of sets itself. We will discuss this in more detail in Chapter 5. For now, let us see how Quine's views on set theory mature and how they fit into and reflect the aims of his philosophical project as a whole.

[90] Ibid., pp. 132–3.
[91] Ibid., p. 133.
[92] See, for example, Quine's "The Inception of 'New Foundations'," p. 288.

CHAPTER 5

Clarifying Our Conceptual Scheme: Set Theory and the Role of Explication

In this chapter, I turn away from Quine's early logical work and the philosophy that I have argued accompanies it. Instead, I consider a central aspect of his more general philosophical project as he describes it in his 1960 *Word and Object*: that of simplifying and clarifying our conceptual scheme.[1] This is not to say that the two philosophical projects are separate undertakings. Indeed, I think they are intimately connected, both in the sense that Quine's basic philosophical outlook emerges from his early work in logic and engagement with Bertrand Russell's work[2] and in the sense that the philosophy of set theory that we saw emerging in Chapter 4 is a natural one to hold in light of the more general philosophical outlook to which Quine is committed, as I will describe in this chapter. In Section 5.1, I begin by sketching the main details of Quine's unique philosophical project of simplifying and clarifying our conceptual scheme, that is, of our current best scientific theory.[3] Central to this undertaking is his appeal to the methods of modern mathematical logic so as to make precise many of the concepts of our science. This will bring me to discuss in Section 5.2 the notion of explication – the clarifying of some unclear expression for a particular purpose, whether scientific or more ordinary[4] – in Quine's philosophy. This will show both how the technical methods of mathematical logic can provide an explication, though an explication does not require such methods, and how his approach to set theory can itself be an explication. In Section 5.3, I turn

[1] Quine, *Word and Object* (Cambridge, MA: MIT Press, 1960).

[2] For more on this point, see my "Quine, Russell, and Naturalism."

[3] The notion of a conceptual scheme may be thought of as having a broader sense than our current best scientific theory does. This is a fair point, but Quine's conception of science is also quite broad; see, for example, W. V. Quine, *From Stimulus to Science* (Cambridge, MA: Harvard University Press, 1995), p. 49. It may, of course, be useful to distinguish between these broader and narrower notions in certain contexts. My account here does not rule this out, but in the rather general account I present here, there will be no occasion to so distinguish between them.

[4] See, for example, Quine, *Word and Object*, pp. 258–9.

to Quine's mature philosophy of set theory as put forth in his 1963 *Set Theory and Its Logic*.[5] Specifically, Section 5.3 situates this work within the context of the broader philosophical aims described for *Word and Object*. Finally, because the position I attribute to Quine bears a striking resemblance to views that are often attributed to Rudolf Carnap, I conclude in Section 5.4 with some analysis of where Quine and Carnap agree and where their important differences lie.

5.1 Clarifying Our Conceptual Scheme

A central aim of Quine's general philosophy is to clarify and simplify our current best scientific theory or, more generally, our conceptual scheme. Certainly, this is a main focus of his 1960 philosophical masterpiece, *Word and Object*. If the presentation there is selective and, in places, schematic, a full and detailed version of such a project is carried out in *Set Theory and Its Logic*, at least with regard to the mathematical portion of our conceptual scheme. In examining this latter undertaking in Section 5.3, we will see an example of what Quine's general philosophy looks like in detail, and we will gain a full account of his mature philosophy of set theory. But let me begin in this section with the relevant details of the more general philosophical project as laid out in *Word and Object*.

Quine begins section 33 of *Word and Object*, "Aims and Claims of Regimentation," by remarking on some of the refinements of language that have recommended themselves earlier in the book, among them the tightening of boundaries for vague terms, clarifying ambiguous terms, and adopting the notation of variables and parentheses to resolve questions of cross-reference and grouping. Such departures from ordinary language, he explains, are in a way also part of ordinary linguistic behavior in that when the difficulties that prompt them persist, the departures themselves become permanent parts of ordinary language. Other such departures are called upon only as needed.[6] In either case, Quine observes that his interests here have dictated departures aimed less at facilitating communication and more at clarifying our conceptual scheme by furthering the understanding of the referential work of our language. While some of these departures have been of a slightly more technical nature – for example, introducing variables and parentheses to resolve ambiguities of cross-reference and grouping – they have all fit

[5] Quine, *Set Theory and Its Logic*, rev. edn (Cambridge, MA: Harvard University Press, 1969).
[6] Quine, *Word and Object*, pp. 157–8.

pretty well within the sorts of ordinary practices of linguistic clarification that are found in everyday discourse.

But there are more technical departures too, in particular in applying logical theory to ordinary language. In this, we find the second aspect of Quine's philosophical project: the simplification of our conceptual scheme. As with clarification, parentheses are a prime example of such simplification, and the benefits that are gained are of no small importance:

> They enable us to iterate a few selfsame constructions as much as we please instead of having continually to vary our idioms in order to keep the grouping straight. They enable us thus to minimize our stock of basic functions, or constructions, and the techniques needed in handling them. They enable us to subject long expressions and short ones to a uniform algorithm, and to argue by substitutions of long expressions for short ones, and vice versa, without readjustments of context. But for parentheses or some alternative convention [fn. omitted] yielding the foregoing benefits, mathematics would not have come far.[7]

The notation of modern logic – what Quine often refers to as "canonical notation" – more generally, in his view, adds much in the way of further simplification to theory. Here, one of the primary contributions is allowing for uniform expression of the variety of expressions available in ordinary language, such as "any" and "every," so that logical theory may be more easily applied:

> Clearly it would be folly to burden logical theory with quirks of usage that we can straighten. It is the part of strategy to keep theory simple where we can, and then, when we want to apply the theory to particular sentences of ordinary language, to transform those sentences into a "canonical form" adapted to the theory. If we were to devise a logic of ordinary language for direct use on sentences as they come, we would have to complicate our rules of inference in sundry and unilluminating ways.[8]

These simplifications by way of a canonical notation better allow for ease of application of logical theory to ordinary language, in particular by allowing for the application of an efficient algorithm of deduction so as to make explicit logical relations of implication among sentences. But this is not all that such a canonical notation makes possible, as Quine goes on to explain:

> [T]he simplification and clarification of logical theory to which a canonical notation contributes is not only algorithmic; it is also conceptual. Each reduction that we make in the variety of constituent constructions needed

[7] Ibid., p. 158.
[8] Ibid., pp. 158–9.

> in building the sentences of science is a simplification in the structure of the inclusive conceptual scheme of science. Each elimination of obscure constructions of notations that we manage to achieve, by paraphrase into more lucid elements, is a clarification of the conceptual scheme of science. The same motives that impel scientists to seek ever simpler and clearer theories adequate to the subject matter of their special sciences are motives for simplification and clarification of the broader framework shared by all the sciences. Here the objective is called philosophical, because of the breadth of the framework concerned; but the motivation is the same. The quest of a simplest, clearest overall pattern of canonical notation is not to be distinguished from a quest of ultimate categories, a limning of the most general traits of reality.[9]

So for Quine, the simplification and clarification of our scientific theories get at the heart of what philosophy has been traditionally thought to aim at: a divulging of the basic features of reality. But as we would expect from Quine, such an activity is not distinct from what science itself aims at. The philosophical pursuit might still be distinguished from that of working scientists but only in its greater generality.[10] Here, we have a prime example of the contribution that Quine thought his logical work could offer to philosophy; both early and late, he aimed at "comprehend [ing] ... the abstract structure of all science."[11]

We have already seen some of how the clarification contributes to a better understanding of the referential aspects of language, but there remains the more general issue of ontology. A particular concern here for Quine is what ontological commitments our conceptual scheme commits us to. He views this as a purely technical endeavor in contrast to the many philosophers who think that our ontology can be discerned through common sense or intuition. Our ontology cannot just be read off of language as it stands, for example, given its tendency to nominalize verbs as a mere stylistic variation.[12] Quine turns again to his canonical notation to address this issue, observing that the objects a theory "admit[s] are precisely the objects which we reckon to the universe of values over which the bound variables of quantification are to be considered to range."[13] The quantifiers "$(\forall x)$" and "$(\exists x)$" in their intended interpretations are read as "every object x such

[9] Ibid., p. 161.

[10] See, for example, Quine, *Word and Object*, pp. 275–6. Hylton (2014) describes Quine's approach as "metaphysics naturalized." Peter Hylton, "Quine's Naturalism Revisited," in *A Companion to W.V.O. Quine*, eds. Gilbert Harman and Ernie Lepore (Malden, MA: Wiley-Blackwell, 2014), pp. 148–62.

[11] Quine, *Time of My Life*, p. 85.

[12] Quine, "Things and Their Place in Theories," in his *Theories and Things*, p. 9.

[13] Quine, *Word and Object*, p. 242.

that" and "there is an object *x* such that," respectively. "The quantifiers," Quine explains, "are encapsulations of these specially selected, unequivocally referential idioms of ordinary language. To paraphrase a sentence into the canonical notation of quantification is, first and foremost, to make its ontic commitment explicit, quantification being a device for talking in general of objects."[14] This criterion for ontological commitment does not, of course, tell us which objects the variables range over, that is, what objects we should allow into our universe for the bound variables of quantification to range over. Quine takes it to be fairly uncontroversial that physical objects will be among the objects of science, but he also includes the abstract objects of mathematics on grounds that science would make very little progress without the aid of mathematics. However, given that all mathematical objects can be rendered in terms of a theory of sets alone, he then limits the universe to include only physical objects and sets.[15]

5.2 Explication

In this section, I want to turn more directly to the notion of explication and its role in Quine's understanding of the philosophical task of simplifying and clarifying our conceptual scheme. This will lead more specifically, in Section 5.3, to an account of Quine's particular approach to the philosophy of set theory as one of explication in his *Set Theory and Its Logic*. It is treating set theory as an explication of the notion of set that serves as the basis for Quine's largely pragmatic, pluralistic, and experimental approach to set theory.

Quine's most sustained discussion of the notion of explication occurs in section 53 of *Word and Object*, "The Ordered Pair as Philosophical Paradigm."[16] While in previous sections he had examined various defective nouns, such as infinitesimals or the ideal objects of physics, and found them to be eliminable as genuine referential fragments of language, here he describes the opposite situation. We have a noun that is

[14] Ibid.

[15] On this point, see chapter 7 of Quine's *Word and Object*. For a more succinct account of deciding on an ontology, see section 6 of his "The Scope and Language of Science," in his *The Ways of Paradox and Other Essays*, rev. and enlarged edn (Cambridge, MA: Harvard University Press, 1976). There is a good deal to say on what Quine admits into his universe, but these issues would take me well beyond my primary concern with set theory. For more on this topic, see Peter Hylton's "Quine on Reference and Ontology," in *The Cambridge Companion to Quine*, ed. Roger Gibson (New York, NY: Cambridge University Press), pp. 115–50, as well as chapter 12 of Hylton's *Quine* (New York, NY: Routledge, 2007).

[16] Quine, *Word and Object* (Cambridge, MA: MIT Press, 1960), sec. 53.

in some sense defective but that is also very useful in its referential status. As Quine explains, "[I]ts utility is found to turn on the admission of denoted objects as values of variables of quantification. In such a case our job is to devise interpretations for it in the term positions where, in its defectiveness, it had not used to occur."[17] We must then somehow make sense of such a noun, preserving its utility while removing its defects. Quine appeals to the ordered pair as a particularly clear case of just this sort of phenomenon.

Typically, we find this device in mathematics, where it allows us to assimilate relations to classes by treating the relations as classes of ordered pairs.[18] Its defectiveness readily appears when we try to give an account of what an ordered pair is. Quoting Charles Peirce's obscure attempt to account for the ordered pair in terms of a mental diagram,[19] Quine recommends instead that "We do better to face the fact that 'ordered pair' is (pending added conventions) a defective noun, not at home in all the questions and answers in which we are accustomed to imbed terms at their full-fledged best."[20] He observes that mathematicians take the single postulate

$$\text{If}\,\langle x, y\rangle = \langle z, w\rangle \quad \text{then} \quad x = z \quad \text{and} \quad y = w, \tag{5.1}$$

to govern all uses required of the ordered pair. So we want a single object that will do the work of two and that satisfies this condition. The solutions, Quine observes, are many, Kazimierz Kuratowski's rendering of $\langle x, y\rangle$ as $\{\{x\}, \{x, y\}\}$ now being among the most common. But Norbert Wiener's $\{\{x\}, \{y, \emptyset\}\}$ serves the purpose equally well. It is straightforward to show that either of these classes satisfies the above postulate.[21] This, Quine declares, is precisely what a philosophical analysis should do:

> This construction is paradigmatic of what we are most typically up to when in a philosophical spirit we offer an "analysis" or "explication" of some hitherto inadequately formulated "idea" or expression. We do not claim synonymy. We do not claim to make clear and explicit what the user of the unclear expression had unconsciously in mind all along. We

[17] Quine, *Word and Object*, p. 257.
[18] Ibid.
[19] Charles Peirce, *Collected Papers*. Vol. II (Cambridge, MA: Harvard University Press, 1932), paragraph 316.
[20] Quine, *Word and Object*, pp. 257–8.
[21] Ibid., pp. 258–9. For a proof of how Kuratowski's definition satisfies the required condition, see for example, Herbert B. Enderton, *Elements of Set Theory* (San Diego, CA: Academic Press, 1977), pp. 35–6. There are plenty of other equally good analyses of the ordered pair; Quine gives further examples on p. 260 of *Word and Object*.

> do not expose hidden meanings, as the words "analysis" and "explication" would suggest; we supply lacks. We fix on the particular functions of the unclear expression that make it worth troubling about, and then devise a substitute, clear and couched in terms to our liking, that fills those functions. Beyond those conditions of partial agreement, dictated by our interests and purposes, any traits of the explicans come under the head of "don't-cares."[22]

The analysis of the ordered pair is unusual only in that the conditions of partial agreement can be made so explicitly and simply. Other cases of analysis will not be so straightforward, but on Quine's account, this is still ultimately what any such analysis is meant to accomplish.[23]

There is no answer, then, to the question of which analysis of the ordered pair is the correct one. Any object satisfying (5.1) has an equal right to being the ordered pair. The decision is largely pragmatic, depending upon the particular context of our needs and aims. We may opt for either of these analyses or even another. For example, if we are working within pure number theory, we might hesitate to introduce an analysis in terms of sets and instead construe $\langle x, y\rangle$ as, for example, $2^x \cdot 3^y$.[24] This is, in general, the situation with regard to analysis. For any such analysis, or explication,

> *explication is elimination*. We have, to begin with, an expression or form of expression that is somehow troublesome. It behaves partly like a term but not enough so, or it is vague in ways that bother us, or it puts kinks in a theory or encourages one or another confusion. But it also serves certain purposes that are not to be abandoned. Then we find a way of accomplishing those same purposes through other channels, using other and less troublesome forms of expression. The old perplexities are resolved.[25]

In the end, any questions about what an ordered pair is are removed when this troublesome notion is replaced by a less troublesome notion. As we have seen, the various explications that have been offered for the

[22] Quine, *Word and Object*, pp. 258–9. "Explication" is, of course, Carnap's terminology; see, for example, his *Meaning and Necessity: A Study in Semantics and Modal Logic*, 2nd enlarged edn (Chicago, IL: University of Chicago Press, 1956), pp. 7–8. Part of what I hoped to have shown in Chapter 4 was that Quine had this notion already in place before any serious engagement with Carnap's work. A more general conclusion, which I have not argued for here but will claim, is that Quine's and Carnap's shared philosophical aims can be traced back to the common influence of Russell.

[23] Here is at least part of his rejection of the analytic/synthetic distinction. Quine just does not think that we have any idea of what the conditions of partial agreement should be for the analysis of analyticity. I will return to this issue in comparing Quine's and Carnap's views on set theory.

[24] There are other options here, as well; see Quine, *Word and Object*, pp. 259–60.

[25] Quine, *Word and Object*, p. 260.

ordered pair conflict with each other, but in Quine's view, there is no need to worry about what to do about the conflicts, for these are among the "don't-cares." This holds generally for explications.[26]

While Quine attributes this notion of an explication to Carnap,[27] the basic idea was already present in Quine's earliest logical work, emerging from his engagement with Russell's logicism. We saw this in Chapter 4 in Quine's rendering of propositions as any "definite technical entity fulfilling just the demands which [we make] of a proposition."[28] These demands of a proposition are what he now identifies as the conditions of partial agreement required by a successful explication. It is notable that Quine brings us back to his earlier engagement with logicism by including the reduction of arithmetic to set theory among his examples of an explication. Here, he presents an explication as a response to the more typically philosophical question "What is a number?" and we have Gottlob Frege replacing these somewhat mysterious entities with entities that are perhaps less questionable, namely, classes (though Quine cites Russell's *Principles of Mathematics* as the source of his remarks on Frege), just as Wiener and Kuratowski did for ordered pairs.[29,30] In this account, for each number *n*, we identify it with the class of all *n*-membered classes (the seeming circularity here can be paraphrased away). To the objection

[26] Elsewhere, Quine makes this point with regard to worries over conflicting explanations. But this is again to confuse explanations with explications, or analyses, in Quine's sense. An explication is an elimination, so we do better to think of alternative eliminations rather than conflicting explanations. On this point, see his *From Stimulus to Science*, pp. 86–7.

[27] Quine, *Word and Object*, p. 259, fn. 4. For Carnap's view of explication, see his *Meaning and Necessity*, pp. 7–8, and his *Logical Foundations of Probability*, 2nd edn (Chicago, IL: University of Chicago Press, 1962), pp. 3–8. For an account of Carnap's views, see Erich Reck, "Carnapian Explication: A Case Study and a Critique," in *Carnap's Ideal of Explication and Naturalism*, ed. Pierre Wagner (New York, NY: Palgrave-Macmillan, 2012), pp. 96–116. For a discussion of how Quine's understanding of an explication diverges from Carnap's, see Martin Gustafsson, "Quine's Conception of Explication – And Why It Isn't Carnap's," in *A Companion to W. V. O. Quine*, eds. Gilbert Harman and Ernie Lepore (Malden, MA: Wiley-Blackwell, 2014), pp. 508–25.

[28] Quine, *A System of Logistic*, p. 33.

[29] It is worth noting that Quine's words are that the classes are "*presumed* to be less in question" [my emphasis] (*Word and Object*, p. 262). This seems to reflect something of Frege's attitude toward classes as being logical objects and therefore more fundamental in some sense to the numbers themselves. In light of the paradoxes, Quine may be acknowledging here that in being a presumption, this claim may not in fact hold true. Still, there are various grounds upon which we could still see this explication as somehow useful. A favorite achievement of the logicist reduction, as Quine sees it, is the unity that it provides to the various branches of mathematics. Thus, by making apparent the interconnections among these branches, such an explication furthers our understanding of arithmetic and mathematics more generally. For an example of this attitude in Quine, see his "Epistemology Naturalized." In W. V. Quine. *Ontological Relativity and Other Essays* (New York, NY: Columbia University Press), p. 70.

[30] Bertrand Russell, *Principles of Mathematics*, 2nd edn (London: George Allen and Unwin Ltd., 1937).

that classes have properties that differ from numbers, Quine responds that this is just to misunderstand the point of explication:

> [N]othing needs be said in rebuttal of those critics, from Peano onward, who have rejected Frege's version because there are things about classes of classes that we have not been prone to say about numbers [fn. omitted]. Nothing, indeed, is more logical than to say that if numbers and classes of classes have different properties then numbers are not classes of classes; but what is overlooked is the point of explication.[31]

Just as with the ordered pair, this is one of many ways of explicating numbers. John Von Neumann and Ernst Zermelo offered other analyses. No two are equivalent, but all serve perfectly well as the numbers. Again, the idea that explication is elimination is central here. There is no question as to which analysis the numbers really are. The numbers have been eliminated in favor of sets. As with ordered pairs, we can provide a condition that any explication of number must satisfy. Such a condition is provided by the notion of a progression given by Giuseppe Peano's axioms for arithmetic, and any objects that satisfy this will serve perfectly well as the numbers.[32]

This attitude has been a constant of Quine's philosophy. We saw it with regard to propositions, where Quine showed how propositions could be rendered in terms of the sequences of his logical system. There was no worry here about whether these are really what propositions are. Sequences of a certain sort turned out to fulfill just the role required of propositions in his system. Quine's point was that there was no further demand to be made of them. We see the same attitude now fully developed and generalized to his philosophy as a whole and so applied also to set theory. This is where his discussion turns next.

I think this is an accurate description of his view, but some qualification is needed. Quine begins the discussion of sets in *Word and Object* by noting that, like the troublesome infinitesimals and the ideal objects of physics, sets are useful to our scientific theorizing. Infinitesimals and ideal objects were found to be dispensable as objects and were ultimately accounted for by other means, though while still preserving the utility they provided to science. We have just seen that they can provide an account of the natural numbers and, by way of the ordered pair, an account of relations. They can be developed further to provide for all of the other number systems of mathematics

[31] Quine, *Word and Object*, p. 262.
[32] Ibid., p. 263.

and of all mathematical objects generally. Indeed, Quine goes on to observe that sets can provide for any abstract object that science demands. But even this might not be enough by itself for accepting the sets into our ontology:

> Such is the power of the notion of class to unify our abstract ontology. To surrender this benefit and face the old abstract objects again in all their primeval disorder would be a wrench, worth making if it were all. But we must remember that the utility of classes is not limited to explication of the various other sorts of abstract objects. *The power of the notion on other counts ... keeps it in continuing demand in mathematics and elsewhere as a working notion in its own right: not only in the protean guises of number, function, state, and all else that it has served to explicate, but also straight. It confers a power that is not known to be available through less objectionable channels* [my emphasis].[33]

So in addition to providing unity, sets also have a place in mathematical research not just because they can be used to give an account of other mathematical objects but also because they also have become, in and of themselves, an integral part of mathematical research. Quine does not give an example of what he has in mind here, but we might consider that the real mathematical interest in Cantor's discovery of set theory was in their providing an account of the mathematics of the infinite. This, we will see in Section 5.3, is where Quine sees the importance of set theory in its own right entering into mathematics.[34]

The utility that sets provide to scientific theory, both as explications and in their own right, cannot then be dismissed lightly. However, Quine described them as being like infinitesimals and ideal objects not just in their utility but also in their troublesome nature. One source of this might be thought to be their abstractness. This is not the real worry, though, since science will always demand abstract objects of some sort, whether they are sets, numbers, or something else. The real source of trouble is the one with which we are already very familiar: Set theory, unconstrained, gives rise to contradictions.[35] Sets as objects, however, cannot be done away with. In this way, the situation is more like that of ordered pairs and numbers. These objects were also found useful but were not altogether dispensable. Despite their unclarity or mysterious nature, their status as objects had to be somehow maintained. And explicating them in terms of sets accomplished this; both ordered pairs and

[33] Ibid., p. 267.

[34] I have in mind here Quine's remark that infinite numbers "are the business of set theory at its most characteristic" (*Set Theory and Its Logic*, p. 5).

[35] Quine, *Word and Object*, p. 266.

the numbers could be eliminated in favor of sets of certain sets. The sets themselves, though, are not eliminated in favor of some other objects that serve as an explication of the sets. The sets are in a sense fundamental to Quine's ontology, though not in any traditionally more metaphysical sense. The sets are ontologically fundamental in that science must adopt some ontology, and the utility and fruitfulness of sets are enough to recommend them. As we saw, Quine also adopts physical objects.

Still, I think explication is the right way to describe Quine's approach to set theory. As I have already remarked, his discussion of set theory follows just after his account of explication here. This is much like his remarks in the slightly later "Ontological Relativity," where he states,

> Expressions are known only by their laws, they laws of concatenation theory, so that any constructs obeying those laws ... are *ipso facto* eligible as explications of expression. Numbers in turn are known only by their laws, the laws of arithmetic, so that nay construct obeying those laws – certain sets, for instance – are eligible in turn as explications of number. Sets in turn are known only by their laws, the laws of set theory.[36]

As in *Word and Object*, sets are not eliminated as numbers might be by an explication, but as with numbers, sets are known only by their laws, so whatever objects satisfy these laws are the sets. Quine's approach to set theory here follows the largely pragmatic one found in the notion of explication. As we have seen, the initial notion of a set must be weakened to the point of consistency while still meeting the conditions of partial agreement for sets. These would be likely to include, for example, preserving some weakened form of the comprehension axiom, that the entities concerned be extensional, and that the theory be strong enough to provide a foundation for mathematics and an account of the mathematics of the infinite. Within these constraints, Quine explains that

> On the whole, however, one manages to salvage most of the utility that the old class theory seemed (in happy ignorance of the paradoxes) to afford, apart from simplicity of governing principles. Naturalness, for whatever it is worth, is of course lost; a multitude of mutually alternative, mutually, incompatible systems of class theory arises, each with only the most bleakly pragmatic claims to attention.[37]

Still, there remains a sense in which Quine's approach *does* eliminate in the way that explications typically do. Recall that an explication serves to

[36] W. V. Quine, "Ontological Relativity," in his *Ontological Relativity and Other Essays* (New York, NY: Columbia University Press, 1969), p. 44.

[37] Quine, *Word and Object*, p. 268.

eliminate "some hitherto inadequately formulated 'idea' or expression."[38] In the case of set theory, this inadequately formulated idea is the naïve conception of set itself; the conception that says every predicate determines a set. This initial thought about sets is eliminated in favor of one of the apparently consistent axiom systems for set theory.

Because the conditions of partial agreement required for an explication do not uniquely determine any one set theory as uniquely correct, set theory for Quine is a largely pluralistic undertaking and so is largely a comparative one as well. This, of course, does not rule out the possibility that some one set theory will eventually triumph over all others. Quine was explicit about this at least as early as his 1941 "Whitehead and the Rise of Modern Logic," as we saw in the quote from this essay included at the beginning of Chapter 4. Even if this does happen, it will be the result of largely pragmatic concerns rather than of someone's having somehow discovered the essence of sets.[39,40] Let us see now how Quine's view of set theory as explication gives rise to the distinctive approach that he develops in his mature philosophy of set theory.

5.3 Quine's Mature Philosophy of Set Theory: *Set Theory and Its Logic*

In Section 5.1, I presented at a fairly general and abstract level an account of how Quine understands the simplification and clarification of our conceptual scheme as a central task for philosophy. As stated at the outset of this chapter, I take it that his mature philosophy of set theory as put forth in his 1963 *Set Theory and Its Logic* is a detailed working out of exactly this sort of philosophical project. We saw in Section 5.2 that an explication need only meet the conditions of partial agreement for a particular concept. Beyond this, there is much freedom to vary the explication. Quine's example of the ordered pair was a prime example of this. For Quine, the notion of set is much the same; it may be explicated in various and conflicting ways depending upon which axioms we choose to adopt. As long as the conditions of partial agreement are satisfied, there

[38] Ibid., p. 258.

[39] Of course, set theory could also remain much like the ordered pair with no one explication ever completely displacing the others. But even if one axiomatization of set theory does eventually win out, this does not show that the others are without interest. They might still have value, say, in certain limited domains, perhaps as classical mechanics does even if we consider relativity to be the more correct and generally applicable theory.

[40] Quine, "Whitehead and the Rise of Modern Logic", in W. V. Quine, *Selected Logic Papers*, enlarged edn (Cambridge, MA: Harvard University Press 1995), pp. 3–36.

is nothing more to demand. If Quine's philosophy of set theory is not focused on finding the uniquely correct account of the notion of set, what, then, does it do? Here is where we can see the philosophical point of *Set Theory and Its Logic*. In line with the philosophical project laid out in *Word and Object* of simplifying and clarifying our conceptual scheme, Quine's philosophy of set theory is a largely a comparative endeavor, aimed at identifying and keeping track of the assumptions that the various axiomatizations of set theory make. It is in this sense that we can view his philosophy of set theory as a full working out of the simplification and clarification of the mathematical portion of our conceptual scheme. Quine aims to keep careful track of at exactly what point we must assume the existence of sets within the mathematical portion of our conceptual scheme and what specific assumptions of mathematical power and ontology the various renderings of set theory make. This is the central task for a Quinean philosophy of set theory.

To this end, Quine adopts his virtual theory of classes[41] as the central device for carrying out this work. This virtual theory allows for the usual class abstract notation "$\{x\colon Fx\}$" for "the class of all *F*s," but it does not allow these purported abstracts to be members of classes or to be quantified over. In this sense, these abstracts are only virtual classes rather than real ones. However, we may still place an object on the left-hand side of the membership relation, $\in$, before a class abstract as in, for example, "$y \in \{x\colon Fx\}$," but this simply reduces to "Fy."[42] In this sense, no real theory of classes is yet in play. The virtual theory is merely a partial counterfeit theory of classes built from pure logic. Only when we allow class abstracts as values of quantifiable variables do we have a real theory of classes. With this move, abstracts may appear on either side of the membership relation.[43] This virtual theory has a number of key benefits for Quine's project. By adhering as much as possible to the virtual theory, he manages to show in a fairly precise manner where exactly mathematics requires the power gained from a real theory of sets. For example, not

[41] Throughout his work on set theory, Quine typically prefers to talk of classes rather than of sets, treating the two terms as basically synonymous. He does not generally mean to draw the technical distinction between sets and classes that is commonly found in the literature, in which sets are those things that can have members and also be members in contrast to classes, or proper classes, which can only have members. He does distinguish carefully between sets and classes when the distinction matters, although he prefers to talk of classes and ultimate classes rather than of sets and (proper) classes (*Set Theory and Its Logic*, pp. 3–4).

[42] We should be careful to distinguish Quine's virtual theory from those set theories that allow for genuine proper, or, as Quine prefers, ultimate, classes. While these theories also prohibit proper classes from being members, they do allow proper classes as values of quantifiable variables. In this sense, such theories are very much theories of real proper classes rather than of virtual ones.

[43] Quine, *Set Theory and Its Logic*, pp. 15–16, 29. I have brushed over some of the more subtle points of the virtual theory; see his section 2 for the full account.

until section 7 does he introduce his first general existence assumptions, allowing for the existence of all classes of fewer than three members, namely, the empty set and pair sets. This allows him to prove indiscriminately that the singleton set $\{\alpha\}$ and the pair set $\{\alpha, \beta\}$ exist for any classes or class abstracts.[44] This also allows him to prove that there are no ultimate, or as is more typical in the literature, proper, classes in his system.[45] Only in section 13 does he make the slightly more extravagant existence assumption of finite classes generally so as to account for mathematical induction. Even here, though, he does not need to posit an infinite class of all finite classes. Rather he introduces an axiom schema (13.1) that assures him of the existence of each individual finite class.[46]

Quine has no general qualms about infinite sets. He will later introduce them, but in keeping with his general approach, he does so only at the point at which mathematics requires this assumption. To make such an assumption to account for the arithmetic of finite numbers seems to him too extravagant if it can be avoided.[47] This is, of course, in contrast with the more usual accounts of arithmetic, which define the natural numbers in terms of the successor operation. For each natural number, there will have to be a successor, and for this, there must be an infinite class. Therefore most set theories assume from the outset an axiom of infinity, as we have seen in both Russell's theory of types and Zermelo's set theory. Quine's trick is to instead define the natural numbers in terms of predecessors so that, while there will have to be arbitrarily large finite numbers, there need be no infinite set.[48] Only when he reaches the laws for the real numbers and the need for a least bound does the demand for infinite sets become unavoidable.[49] Quine then does not hesitate to make this assumption. After all, it is the concern with infinite classes and their relative sizes that Quine describes as the real core of set theory.[50] The point is only to keep track of assumptions, knowing that we can make them, and to let them into our system as needed.

[44] Under the virtual theory, we might have class abstracts to which no real class corresponds.

[45] Quine, *Set Theory and Its Logic*, pp. 50–2.

[46] Ibid., pp. 87–8.

[47] Recall that we saw this attitude expressed already in his 1938 paper, "On the Theory of Types."

[48] Quine, *Set Theory and Its Logic*, pp. 75–6.

[49] Ibid., p. xii, Ch. 6.

[50] Ibid., p. 5. Quine notes of his theory of ordinals that they are von Neumann's version whereas his natural numbers are Zermelo's. A single account of both might seem a better option, but Quine disagrees, observing that his approach allows us to "preserve economy of assumption at the level of natural-number theory; and so I let the natural numbers and the ordinals go their separate ways" (*Set Theory and Its Logic*, pp. xii–xiii). Later in the work, he approvingly cites Zermelo's adoption of an axiom of infinity so as to provide an account of the real numbers and whatever mathematics lies beyond them (*Set Theory and Its Logic*, p. 282).

Aside from minimizing and postponing ontological assumptions, Quine's virtual theory has a second but related aim. Because part of the clarificatory aspect of his project is the comparison of the various and largely incompatible axiomatizations of set theory with their differing assumptions of power and ontology, Quine needs some fairly neutral ground upon which he can place this discussion. No one of these axiomatizations can serve this purpose, since it would privilege one of them over the others. He initially suggests that we might try a somewhat informal approach to this issue. He dismisses this attempt on grounds that very little serious reasoning in set theory can be done without making some substantial assumptions and that these are the kinds of assumptions that could lead us to the various set-theoretic paradoxes. We might try just talking around the paradoxes, but this raises a difficulty for the novices in set theory who have already heard of them. They will be left wondering which intuitive arguments will still hold up in light of the paradoxes. So Quine intends his virtual theory as the basis for his desired neutral starting point:

> I handle the problem by hewing a formal line from the outset, but keeping my axioms weak and reasonable and thus nearly neutral. I postpone as best I can the topics that depend on stronger axioms; and, when these topics have to be faced, I still postpone the stronger axioms by incorporating necessary assumptions rather as explicit hypotheses into the theorems that require them. In this way I manage to introduce the reader at some length to the substance of set theory without any grave breach of neutrality, and yet also without resorting to studied informality or artificial protraction of innocence.[51]

Only after proceeding in this way for ten chapters does he turn to consider and compare the more standard axiomatizations of set theory.

This culminating part of the work is the real core of Quine's philosophy of set theory as an exploratory and experimental undertaking. Here, he compares and contrasts the four main systems of set theory that are commonly found in the literature –Russell's type theory, Zermelo's system, Quine's own NF and ML, and von Neumann's system[52] – and their respective and often incompatible assumptions. As we have seen, the two initial axiomatizations of set theory, Russell's and Zermelo's, both came in 1908. Quine begins with Russell's and its development in

[51] Quine, *Set Theory and Its Logic*, pp. x–xi.

[52] I will not discuss ML and von Neumann's theories here, as they turn the discussion toward set theories with classes added, a topic that is not specifically relevant to my immediate concerns.

response to the paradoxes. We have already seen an account of Russell's theory in Chapter 2, so I will be brief with Quine's account of it and just touch on of some of its key features. Quine begins with Russell's attempt, following Henri Poincaré, to diagnose the paradoxes as resulting from a single source, that of a sort of vicious circle. As is familiar by now, Russell's paradox results from specifying a class y by way of non-self-membership, that is, by the condition $(x)(x \in y \equiv x \notin x)$, and then letting the quantified variable x take y itself as a value. The basic thought here behind diagnosing the paradoxes was that it is somehow illegitimate to presuppose y in the range of the quantifier when specifying y itself.[53]

Quine finds nothing especially vicious in this approach when we think of the classes as being there from the start with particular ones only standing in need of specification (and according to Quine, Poincaré suggests no temporal aspect to the creation of classes). For example, the most typical Yale man is specified on the basis of the average Yale scores including his own. So while there is nothing obviously viciously circular in impredicative specification of classes, Quine does see impredicativity as pointing toward one way of restricting the law of comprehension so as to pare down the universe of classes to consistency. While carefully noting that it is only a metaphor, he does point to this proposal as being less arbitrary than others in the constructional metaphor it suggests; the classes are just those that could be generated from unspecified beginnings over an infinite period of time by way of a membership condition mentioning only classes previously generated. We saw in Chapter 2 the details of how Russell carries this out in terms of his theory of types and orders. Leaving the constructional metaphor aside, this set theory has as its characteristic feature that its universe allows a transfinite ordering of its classes in that the membership condition for each class can be given only in terms of classes occurring earlier in the ordering.[54] Again, though, Russell's initial theory of orders is an intensional theory of propositional functions, not an extensional theory of classes. Once he applies his axiom of reducibility and contextually defines the classes in terms of propositional functions, the original intensional theory becomes, in a sense, superfluous. As we saw in his earliest work in set theory, with the

[53] Quine, *Set Theory and Its Logic*, p. 242.

[54] Ibid., pp. 242–3. Quine goes on to describe the details of Russell's theory in more detail, beginning with his intensional basis for classes and then transitioning to the modern extensional theory of types. Because we have already seen the relevant details in Chapter 2, I will here skip over most of Quine's discussion on these matters.

purpose of furthering clarity, Quine recommends dropping the intensional theory altogether.[55]

While the hierarchy of types does serve to block the paradoxes, Quine observes that the restriction imposed on the comprehension axiom is of a rather drastic sort in that it forces a revision of logic in that type theory is a many-sorted logic. As we have seen in various places, Russell frequently passes over the specific types of the variables in favor of what he calls typical ambiguity, a suppressing of the type indexes on the variables of a given formula. For example, appealing to the practice of typical ambiguity, we write the formula "$(\exists y)(x)(x \in y)$," but this is not itself a formula of type theory. It is rather like "$(\exists y^{n+1})(x^{n})(x^{n} \in y^{n+1})$," where n is schematic for some specific type index. As an actual formula of type theory, this might be one of "$(\exists y^{1})(x^{0})(x^{0} \in y^{1})$," "$(\exists y^{2})(x^{1})(x^{1} \in y^{2})$," and so on. The schematic version just stands in place of an infinite number of actual formulas of type theory. So despite the suppressing of type indexes, the formula remains one of a many-sorted logic rather than of the familiar logic of truth-functions and quantifiers. Quine's maxim of minimum mutilation cautions against disrupting general logic just to fend off the paradoxes, which are only a part of the special subject matter of set theory.[56] This brings Quine to consider Zermelo's set theory with its general variables.

Quine observes that instead of providing each variable with a specific type index, we might just have one sort of variable and introduce a predicate, "T_n," to indicate the various types.[57] So, for example, the formulas "$(x^n)Fx^n$" and "$(\exists x^n)Fx^n$" of type theory can now instead be written with general variables, restricted in the usual way, as "$(x)(T_n x \supset Fx)$" or "$(\exists x)(T_n x \, . \, Fx)$." With general variables restored, Russell's grammatical restriction disappears. Whereas in Russell's system, variables on the left- and right-hand sides of the membership relation had to be of consecutive

[55] Quine has made this point on a number of occasions and has criticized Russell for adopting the intensional theory so as to then define the less obscure in terms of the more obscure. Quine himself observes that only the foundational portion, roughly the first 200 pages, of *Principia Mathematica* appeals to the intensional theory of propositional functions. The rest is carried out in terms of classes and relations-in-extension. As we saw earlier in this chapter, this is exactly what we should expect from Russell: Russell thought the propositional functions to have more claim to being logic than the theory of classes had. This is not to say that Quine's criticisms miss their mark, but we should also not ignore that Russell thought the intensional basis to be better motivated for his purposes than the extensional one that Quine urges. Quine's account of Russell's move from propositional functions to classes is on pp. 250–2 of *Set Theory and Its Logic*.

[56] Quine, *Set Theory and Its Logic*, pp. 266–7.

[57] We will further consider this move from type theory to Zermelo's theory in Chapter 7.

ascending types to be meaningful, formulas that violate this restriction are now simply false. Having done away with the restriction, Quine asks whether Russell ever needed it at all, responding that

> he did not. One sees how he was led to make it. The idea of getting the paradoxes to turn out neither true nor false but meaningless was an inviting one, and the failure of definition through vicious circularity was of a piece with it. But his system itself, by the time he had it working, was working independently of these ideas.[58]

This, as we will see, is an important theme for Quine and one that he frequently returned to, particularly in his development of NF.

Restricting the general variables by way of predicates for specific types allows for the translation of all theorems of types with special variables into the new theory. But Quine observes that this new system "is more glaringly incomplete than the former, because of all the newly meaningful formulas whose proof or disproof we have not provided for."[59] Among these that are clearly valid for types are "$x \in y \supset T_n x \equiv T_{n+1} y$," "$(\exists y)(x)(x \in y \equiv x \in z\ .\ Fx)$," and "$(\exists y)(x)(x \in y \equiv (\exists w)(x \in w \in z))$," all three of which Quine says we might add as axioms or an axiom schemata. The latter two have exactly this status in Zermelo's system as his axiom schema of separation and the union axiom. In addition to being led to Zermelo's system by adding these additional axioms, Quine observes additional simplifications that are suggested by the change to general variables, such as equating the null classes for each class type.[60]

To gain further insight into set theory, Quine takes a comparative approach and proposes looking at how Zermelo's theory addresses Russell's paradox. As we saw, Russell's theory avoids his paradox by preventing self-membership. Zermelo's theory, with its general variables, has no such typing restrictions; instead, it restricts comprehension by way of the separation schema "$(\exists y)(x)(x \in y \equiv x \in z\ .\ Fx)$" mentioned in the previous paragraph. Separation allows only for specification of a set as a subset of a set z. The trick to getting Russell's paradox is to find a z big enough that we may specify a set of all x such that x is not a member of itself. But Quine observes that there seems to be no way of doing so:

[58] Quine, *Set Theory and Its Logic*, p. 269.
[59] Ibid., p. 271.
[60] Ibid., pp. 272–3.

"[W]e need a *z* big enough that $\{x: x \notin x\} \subseteq z'$[sic], and there is no visible way to such a *z*. Zermelo's protection against the paradoxes consists essentially in eschewing too big classes."[61] We will consider Zermelo's system further in Chapter 6. For now, I should emphasize that while Quine has brought out differences between Russell's type theory and what will become Zermelo's theory, he has importantly stressed the rather natural connections between the theories. Even in discussing their distinct approaches to resolving Russell's paradox, his overall aim is better understanding of set theory generally. Whereas much contemporary philosophy of set theory aims to privilege one axiomatization over another, Quine's approach is one of tolerance in the interest of furthering our understanding of set theory more generally.

We have already given substantial consideration in Chapters 3 and 4 to Quine's own set theory, NF, and we will add to this discussion in Chapters 6 and 7. Let me add to this matter here as well. In light of our most recent discussions of Russell's theory of types and Zermelo's of cumulative types, let me comment on the relevance of this aspect of set theory to NF. We have seen how naturally type theory can be accounted for with general variables and then how easily types can then be transformed into Zermelo's theory. In a way, what this transformation shows is that there is no inherent need for the set-theoretic universe to be typed into levels. One way of looking at NF, already suggested by what Quine identifies as the formal restriction of type theory, is that NF represents a sort of minimal assumption of the supposed set-theoretic hierarchy, in contrast to that found in Russell's types and in Zermelo's cumulative types. Like Zermelo's theory, NF's variables are completely general, but Zermelo's theory still holds on, in its own way, to the idea of the set-theoretic universe being divided into levels. In this way, unlike Quine's NF, Zermelo's theory does not completely abandon the ontological doctrine found in type theory. Quine's requirement of stratification is what is left behind when this metaphysical picture of the set-theoretic hierarchy is dropped. As Quine remarks, "In NF there are no types. Nor is it required that formulas be stratified to be meaningful. Stratification is simply an ultimate, irreducible stipulation to which a formula is to conform *if* it is to qualify as a case of '*Fx*' in the particular axiom schema [of comprehension] [Quine's italics]."[62] Again, Quine does not ignore differences, but in drawing them out, he shows how much overlap there is

61 Ibid., pp. 277–8. Hallett's *Cantorian Set Theory and Limitation of Size* is the standard source for more on the history and philosophy of the limitation of size approach to set theory.

62 Quine, *Set Theory and Its Logic*, p. 289.

between the various set theories. As much as Quine wants to stress where these alternative axiomatizations diverge, he also goes to great lengths to bring out their structural connections.

We might find Quine's open-mindedness about set theory surprising, given his usual preference for ontological parsimony and his oft-stated commitment to classical first-order logic. Let me comment on the latter point first. I take it that Quine thinks logic a much more settled and uncontroversial science than set theory is. Revising logic, while in principle always an option, would be a far more drastic change to our conceptual scheme than is moving from one set theory to another. A change in logic violates Quine's maxim of minimum mutilation in a way that a change in set theory does not, especially given Quine's view that set theory is still a far from settled discipline.[63] As to ontological parsimony, this is a preference that Quine has often expressed, but just as with his commitment to classical first-order logic, it is not an unexceptionable dogma of his philosophy. Indeed, it is much less of a commitment than the one to classical first-order logic. Quine does think that we should not make unnecessary ontological assumptions, but we also cannot rule out from the start which assumptions to accept and which to reject. This will be an ongoing process of experimentation and exploration of just the sort that Quine presents in *Set Theory and Its Logic*. As we have now seen, this process focuses largely on the investigation of the various assumptions of strength and ontology found in the different axiomatizations of set theory. This approach furthers our understanding of set theory as a whole:

> It is generally conducive to understanding to keep track of our presuppositions, in point of objects and otherwise, project by project; and to welcome ontological economy in connection with one project even if a more lavish ontology is needed for the next. But it is also important to have the less economical and more powerful mathematical theories well in hand as engines of discovery, for swift use in unforeseen places – even though in each such case we take pains afterward to find more economical ways of gaining the same result.[64]

[63] This in itself may be a controversial claim about set theory, given the now wide acceptance of Zermelo's set theory. The claim would have been much less controversial during the period in which Quine did the bulk of his work in set theory. But we might also question whether matters are as settled as they now seem to be with regard to set theory. This will be discussed further in Chapters 6 and 7. For a recent account of Quine's views on logic with regard to revisability, see Matthew Carlson, "Logic and the Structure of the Web of Belief," *Journal for the History of Analytical Philosophy* 3:5 (2015): https://jhaponline.org/jhap/article/view/28.

[64] Quine, *Word and Object*, p. 270.

Understanding is, after all, for Quine, one of the central aims of science and so also of philosophy.[65]

5.4 An Addendum on Quine and Carnap on Tolerance and Set Theory

As we have seen, Quine states explicitly his agreement with Carnap over the notion of explication.[66] Quine urges tolerance in his set theory, taking an exploratory and experimental approach to the subject, which might also suggest agreement with Carnap.[67] However, in his debates with Carnap over analyticity, Quine has typically been understood as rejecting Carnap's famed principle of tolerance.[68] In this final section, I want to explore more carefully the sense in which Quine adheres to a version of tolerance and how it differs from Carnap's. While bringing out their differences, however, I would also like to emphasize that Quine and Carnap agree that the fundamental approach to philosophy should be a scientific one. It is precisely their different understanding of how this should be done in such a way that the fruitless metaphysical debates of the past are overcome that leads Carnap to embrace the principle and Quine to reject it. Let me begin with Carnap.

As is well known, much of Carnap's work was devoted to reshaping philosophy to bring it into line with the sciences so that it might avoid the "empty wrangling" of the past.[69] He saw this as particularly evident in the debates over the foundations of mathematics that arose in the 1920s between logicism, intuitionism, and formalism. Rather than taking a position as to whether any one of these views is correct, Carnap saw value in all of them and thought that dogmatic adherence to any one of them was contrary to the nature of science.[70] It is in this spirit that he urged his principle of tolerance in his 1934 *Logical Syntax of Language*:

> To eliminate this standpoint [that there is a single correct language logic], together with the pseudo-problems and wearisome controversies

[65] Quine, *Pursuit of Truth*, p. 2.

[66] Whether this is an accurate characterization might be questioned, as it is by Gustafsson in his "Quine's Conception of Explication – And Why It Isn't Carnap's."

[67] We will see that Carnap himself thought this, describing their differing views on ontology as really just a terminological difference.

[68] Both Thomas Ricketts and Peter Hylton see tolerance as being at the heart of Quine and Carnap's dispute over the analytic/synthetic distinction. See, for example, Thomas Ricketts, "Languages and Calculi," in *Logical Empiricism in North America*, eds. Gary L. Hardcastle and Alan W. Richardson (Minneapolis, MN: University of Minnesota Press, 2003), pp. 257–80, and Hylton, *Quine* , pp. 72–3.

[69] Rudolf Carnap, *Philosophy and Logical Syntax*, London: Thoemmes Press, 1996 [originally published 1935], p. 81.

[70] For some indication of this view see Rudolf Carnap, "Intellectual Autobiography," in P. A. Schilpp, ed., *The Philosophy of Rudolf Carnap* (Lasalle: Open Court, 1963), pp. 3–84. For his discussion of the main philosophies of mathematics, see pp. 46–50; for his general attitude of scientific approach to philosophy, see pp. 18–19.

> which arise as a result of it, is one the chief tasks of this book. In it, the view will be maintained that we have in every respect complete liberty with regard to the forms of language; that both the forms of construction for sentences and the rules of transformation ... may be chosen quite arbitrarily.[71]

The idea that is encapsulated in Carnap's principle is that there is no single correct language of science. As he dramatically explains later in the book, "*In logic, there are no morals.* Everyone is at liberty to build up his own logic, i.e. his own form of language, as he wishes. All that is required of him is that, if he wishes to discuss it, he must state his methods clearly, and give syntactical rules instead of philosophical arguments" [Carnap's italics].[72] Previously, languages had been constructed by first assigning meanings to the primitive logical and mathematical symbols. This procedure determined the logical correctness of the sentences and inferences of the system of language. However, Carnap explains that since such meanings are given in language and so are inexact, any consequences following from them will be similarly inexact. In accord with his principle of tolerance, he instead urges that we choose a set of postulates and rules arbitrarily. This choice then will determine the meaning assigned to the primitives of the system. This method will have particular bearing on the disputes over the foundations of mathematics:

> By this method, also, the conflict between the divergent points of view on the problem of the foundations of mathematics disappears. For language, in its mathematical form, can be constructed according to the preferences of any one of the points of view represented; so that no question of justification arises at all but only the question of the syntactical consequences to which one or other of the choices leads, including the question of non-contradiction.[73]

So we have complete freedom to develop any language system we like. Furthermore, since the adoption of the postulates and rules then determine logical correctness of the system's sentences and inferences, there is no question as to whether the adoption of the system itself is right or wrong. Such questions can arise only after a language system – or what Carnap would later call a linguistic framework – has been adopted.

The general philosophical significance of Carnap's standpoint emerges perhaps most clearly in his 1950 "Empiricism, Semantics, and

[71] Rudolf Carnap, *Logical Syntax of Language*, trans. Amethe Smeaton (Paterson, NJ: Littlefield, Adams, and Company, 1959), pp. xiv–xv.
[72] Ibid., p. 52.
[73] Ibid., p. xv.

Ontology."[74] It is here that Carnap elaborates his principle and its consequences in terms of the distinction between internal and external questions. He sets his discussion within the general context of debates between realists and nominalists over whether abstract objects should be allowed into our scientific theories.[75] Empiricists have tended to favor the nominalist position of avoiding such entities to the greatest extent possible, but science cannot get very far without some abstract objects, specifically the objects of mathematics. Carnap's aim is to clarify this dispute and to show how scientific philosophers can adopt abstract objects without crossing over into the more questionable realm of metaphysics. As he explains, "[I]t will be shown that using such a language does not imply embracing a Platonic ontology but is perfectly compatible with empiricism and strictly scientific thinking."[76] It is to demonstrate this point that he introduces his talk of linguistic frameworks and the distinction between internal and external questions.

To construct a linguistic framework, Carnap describes a situation in which we may want to speak of a new kind of entities in our language. These new entities will require a new way of speaking, so we will have to introduce a set of rules that govern how we may speak of them. For example, suppose we wanted to introduce numbers into a language that did not include them previously. We would have to have a set of rules governing our talk of numbers. This would include a set of axioms, such as the Peano axioms for arithmetic, and a set of rules for logical inferences from these axioms. This would constitute a linguistic framework for numbers. Given a linguistic framework, two sorts of questions can arise: internal and external questions. Internal questions are formulated within a linguistic framework and concern the entities of that framework. In the linguistic framework for numbers, we might ask whether there are prime numbers greater than a million and attempt to find and formulate an answer also within this framework. Similarly, within a linguistic framework for physical objects, we might ask where a particular object is located and then attempt to answer this question using the means provided by that framework.[77] In both of these cases, these questions are language relative. We can ask about numbers in the former case because we have adopted a linguistic framework for numbers, and we can ask

[74] Rudolf Carnap, "Empiricism, Semantics, and Ontology," in *Meaning and Necessity*, 2nd edn (Chicago, IL: University of Chicago Press, 1956).

[75] More specifically, the concern is with whether abstract objects such as propositions and properties should be accepted into a semantic theory.

[76] Carnap, "Empiricism, Semantics, and Ontology," p. 206.

[77] Carnap discusses linguistic frameworks and gives various examples on pp. 206–13 of "Empiricism, Semantics, and Ontology."

about physical objects in the latter case because we have adopted a linguistic framework for physical objects. In either case, there is no question of absolute existence of the entities involved. Questions of existence will always depend on what language we have adopted.

But there are still questions that we might want to ask about the reality of the system of new entities as a whole. Carnap describes this sort of inquiry as what many philosophers have meant by an ontological question: "Many philosophers regard a question of this kind as an ontological question which must be raised and answered *before* the introduction of the new language forms. The latter introduction, they believe, is legitimate only if it can be justified by an ontological insight supplying an affirmative answer to the question of reality" [Carnap's italics].[78] Such questions as these, because they take place before a particular linguistic framework has been adopted, are the external questions. Carnap rejects this more metaphysical approach to philosophy; since they apparently take place outside of any linguistic framework, they are nonsensical. But by appealing to his linguistic frameworks, he reconstrues these traditional questions of ontology so as to render them free of metaphysics:

> We may still speak … of "the acceptance of the new entities" since this form of speech is customary; but one must keep in mind that this phrase does not mean for us anything more than acceptance of the new framework, i.e., of the new linguistic forms. Above all, it must not be interpreted as referring to an assumption, belief, or assertion of "the reality of the entities." There is no such assertion. An alleged statement of the reality of the system of entities is a pseudo-statement without cognitive content.[79]

However, Carnap's reconstrual still leaves these questions about the system as a whole outside of the system itself, so what are we to say of them from his scientific approach to philosophy? Carnap takes such questions to be in principle different from the internal, or theoretical, questions, as he next explains:

> To be sure, we have to face at this point an important question; but it is a practical, not a theoretical question; it is the question of whether or not to accept the new linguistic forms. The acceptance cannot be judged as being either true or false because it is not an assertion. It can only be judged as being more less expedient, fruitful, conducive to the aim for which the language is intended. Judgments of this kind supply the motivation for the decision of accepting or rejecting the kind of entities.[80]

[78] Carnap, "Empiricism, Semantics, and Ontology," p. 214.
[79] Ibid.
[80] Ibid.

So the external questions are pragmatic in nature rather than theoretical. They cannot be true or false because questions of truth or falsity are relative to a particular linguistic framework. Because the framework must be adopted before such questions can arise, it makes no sense, from Carnap's perspective, to judge the adoption of a particular framework as correct or incorrect. We can urge the adoption of one framework over another only on pragmatic grounds such as simplicity, elegance, or fruitfulness.

As I noted above, Carnap took such an approach to avoid the unprogressive disputes of past philosophy. Whereas previous philosophers appeared to reach insoluble standoffs between their various and incompatible views, Carnap's method is supposed to show the way out of these insoluble disputes by exposing the pseudo-problems that underlie them. There is no question of whether, for example, numbers really exist in some absolute metaphysical sense. There is only the question of which linguistic framework to adopt. With the choice of the framework made explicit, there is no longer any genuine dispute to be resolved. To take an example from the foundations of mathematics, we might adopt the comparatively weak mathematical language of intuitionism in one setting and the comparatively strong one of classical mathematics in another, as was Carnap's own approach in his *Logical Syntax of Language*. Depending upon our aims, the safety of the former might be preferable to the greater ease of use and simplicity of the latter or vice versa.[81] More generally, we might say the same of disputes between nominalists and realists. Again, there is no real dispute here, only a difference of linguistic frameworks, and either might be chosen depending upon which serves our purpose better. Overcoming these pseudo-debates with the goal of making philosophy a truly scientific and progressive discipline is at the heart of Carnap's urging of tolerance, as he concludes,

> The acceptance or rejection of abstract linguistic forms ... will finally be decided by their efficiency as instruments, the ratio of the results achieved to the amount and complexity of the efforts required. To decree dogmatic prohibitions of certain linguistic forms instead of testing them by their success or failure in practical use, is worse than futile; it is positively harmful because it may obstruct scientific progress ... Let us grant to those who work in any special field of investigation the freedom

[81] Carnap suggests these considerations explicitly in his *Foundations of Logic and Mathematics* (Chicago, IL: University of Chicago Press, 1939), pp. 192–3. Quine also suggests them in his *Set Theory and Its Logic*, in the final section comparing the various systems of set theory (pp. 328–9).

> to use any form of expression which seems useful to them; the work in the field will sooner or later lead to the elimination of those forms which have no useful function. *Let us be cautious in making assertions and critical in examining them, but tolerant in permitting linguistic forms* [Carnap's italics].[82]

On the face of it, Quine's approach to set theory might seem very much like Carnap's general attitude of tolerance. Indeed, Carnap himself observes – citing Quine's remark that when pursuing ontology "the obvious counsel is tolerance and an experimental spirit"[83] – that their divergence over ontology is really only a terminological one.[84] But is this really so? Many commentators have argued that Quine's rejection of Carnap's principle of tolerance is central to their differences over the legitimacy of the analytic/synthetic distinction.[85] Perhaps Quine takes a different attitude toward set theory than he does toward other scientific theories. We will see that there may be some truth in understanding Quine in this way, but on the whole, I think that his attitude toward tolerance is the same for all scientific theories. After all, we just saw that he urges tolerance in the case of ontology generally. Let us return now to Quine so that we may see more clearly the sense in which Quine can urge tolerance while still rejecting Carnap's principle of tolerance.

Throughout our discussions of Quine's philosophy of set theory, we have seen him urging a tolerant attitude, and this is true of both his early and late views. Recall, for instance, his remark in the 1941 paper "Whitehead and the Rise of Modern Logic" that none of the various set theories "has an intuitive foundation," so the best among them will be that which "engenders a form of logic most convenient for mathematics and the sciences."[86] As we have seen, this attitude carries through into Quine's mature philosophy of set theory with its emphasis on exploration and experimentation and the comparison of the various set theories. So we also cannot explain Quine's tolerance by saying that it was an early

[82] Carnap, "Empiricism, Semantic, and Ontology," p. 221. This attitude is also evident in his earlier *Logical Syntax of Language* and *Foundations of Logic and Mathematics*, although in those works, his focus is specifically on the foundations of mathematics. In his "Autobiography," Carnap makes the point that some version of this view was always present in his thinking. Carnap, "Intellectual Autobiography." In Schilpp, P. A. *The Philosophy of Rudolf Carnap* ((La Salle, IL: Open Court, 1963), pp. 3–84.

[83] Quine, "On What There Is," p. 19.

[84] Carnap, "Empiricism, Semantics, and Ontology," p. 215, fn. 5.

[85] For some examples, see fn. 62.

[86] Quine, "Whitehead and the Rise of Modern Logic," p. 27.

commitment that he shifted away from in his later work.[87] How, then, does he understand his notion of tolerance?

As we saw, Carnap's version of tolerance is directly connected to his notion of a linguistic framework, which allows him to render philosophical questions free of metaphysics by reconstruing them as questions concerning only of which linguistic framework to adopt. The crucial distinction for Carnap here is that the choice of a linguistic framework is in principle different from the theoretical claims made within that framework. The former can be evaluated only by pragmatic criteria such as expedience, fruitfulness, and whether it suits the aims for which the linguistic framework was intended.[88] The latter, however, are evaluated in accordance with the logical and empirical methods provided by that framework and so can be judged true or false. We should be tolerant toward the framework because there can be no evaluation of it as correct or incorrect, since justification is always relative to a particular linguistic framework. Here is precisely the point at which Quine rejects the kind of tolerance that Carnap urges. Instead, in Quine's view, in both cases we are appealing to the pragmatic criteria that Carnap says are alone operative in the choice of a linguistic framework.[89] Quine's holism is crucial here, that is, his view that a statement is to be evaluated not by direct comparison with experience but rather by looking at its contribution to the theory as a whole.[90] For Quine, we are

[87] However, Quine's sympathy with Gödel's constructible universe, $V = L$, may indicate a growing preference for ZF style set theory in his later work. See, for example, Quine's 1990 "Immanence and Validity" reprinted in *Selected Logic Papers*, p. 243, or his *Pursuit of Truth*, p. 95. Such a preference is in no way inconsistent with Quine's exploratory and experimental approach. After all, part of ongoing science is that one theory may emerge as dominant and so be taken as correct. Quine never indicates that he thinks we have reached this point with set theory, but there is no reason to rule out that we might be heading in that direction.

[88] Carnap, "Empiricism, Semantics, and Ontology," p. 214.

[89] The discussion that follows in this paragraph follows Peter Hylton's exposition in his *Quine*, pp. 71–3, as well as Hylton's "Analyticity and the Indeterminacy of Translation," *Synthese* 52:2 (1982), pp. 173–5. In the latter, Hylton notes a possible response from Carnap on this point: He might counter that only pragmatic criteria are operative in choosing a framework, whereas both confirmational and pragmatic criteria are operative in assessing theoretical claims. This would still mark a difference, in principle, between the two, since assessing the theoretical claims would still be a rule governed activity, that is, something that fulfills Carnap's ideal of rational adjudicability in contrast to the choice of framework. From Quine's perspective, however, this begs the question, since it is the status of such rules that are in question. This more complex view seems more likely to be Carnap's actual view, since he explicitly admits that a version of holism is operative within a linguistic framework as well. See, for example, his *Logical Syntax of Language*, pp. 317–18.

[90] This does not mean that we need to evaluate the sentence in terms of its contribution to the whole of science. We may look at its contribution to some subtheory. On this point, see W. V. Quine, "Two Dogmas in Retrospect," reprinted in W. V. Quine, *Confessions of a Confirmed Extensionalist and Other Essays*, eds. Dagfinn Føllesdal and Douglas B. Quine (Cambridge, MA: Harvard University Press, 2008), p. 393.

assessing how well an entire theory allows us to cope with our sensory experience. Thus the most abstract claims of science and mathematics, which Carnap typically renders to the framework for science, are evaluated in the same way as are the claims that are more directly related to sensory evidence. That is, we accept the individual claims of a theory according to whether they are part of a theory that handles experience overall better than its rivals do. The theory as a whole is evaluated in terms of such features as elegance, simplicity, and fruitfulness – just the sort of criteria that Carnap said are operative at the level of choosing a linguistic framework. So for Quine, in contrast to Carnap, there is no real difference between how we evaluate the choice of the linguistic framework and how we evaluate the choice of a theory within that framework. Justification, then, is not a language-relative matter.

What of set theory more specifically? If Quine rejects Carnap's principle of tolerance, on what grounds can he urge tolerance in set theory?[91] In his 1954 "Carnap and Logical Truth," Quine directly addresses set theory in his arguments against Carnap. Here, Quine says that set theory might be accurately described as being true by linguistic convention, suggesting again that his view is much the same as Carnap's.[92] But we see Quine immediately emphasizing that he treats set theory as interpreted mathematics and so as being very much about sets:

> In set theory we discourse about certain immaterial entities, real or erroneously alleged, viz., sets, or classes. And it is in the effort to make up our minds about genuine truth and falsity of sentences about these objects that we find ourselves engaged in something very like convention in an ordinary non-metaphorical sense of the word. We find ourselves making deliberate choices and setting them forth unaccompanied by any attempt at justification other than in terms of elegance and convenience. These adoptions, called postulates, and their logical consequences …, are true until further notice.[93]

Here, we see already how Quine's position differs from Carnap's. Recall that, in accordance with his principle of tolerance, Carnap thought we should have the complete freedom to develop a variety of linguistic frameworks and that we should do so in the pursuit of scientific progress. Furthermore, Carnap thought that his linguistic frameworks freed him from

[91] As I remarked above, whatever tolerance Quine urges in set theory is one that he will urge for science as a whole.

[92] W. V. Quine, "Carnap and Logical Truth," in *The Ways of Paradox*, p. 115. In his "Autobiography," Carnap say that the principle of tolerance might be more accurately described as the "principle of conventionality of language forms" (p. 55).

[93] Quine, "Carnap and Logical Truth," p. 117.

the metaphysics of the past. In adopting a linguistic framework, we were only adopting a way of speaking, not committing ourselves to an ontology in any robust sense. Quine does not share such worries. By focusing on interpreted theories, he immediately shows his willingness to engage in ontology. Furthermore, our conventional decisions to treat certain claims about these entities as true show that these conventions are not prior to our theorizing about sets; this is not a prior decision to adopt a particular language form. Quine's conventions are a part of the very science of set theory.

Quine goes on to distinguish two types of postulation in which we might engage here: discursive and legislative. The former is merely the selecting of some truths as postulates from among the existing body of truths concerning the relevant entities. This sort of postulation just imposes some ordering of the truths belonging to a given theory. We might think of this as what goes on with Euclidean geometry. We have the truths, and then we select some of them as the postulates from which the others may be derived. The latter sort of postulation is what we have here in set theory. This type of postulation takes some claim about the entities concerned – in this case, sets – and declares our postulate to be true of them. In this case, postulation is the source of truth. Whereas Carnap took this latter kind of postulation to mark a significant trait in distinguishing external from internal questions, Quine does not. Postulation, whether legislative or discursive, refers simply to that particular act of postulation and has no enduring consequence. "So conceived," he explains, "conventionality is a passing trait, significant at the moving front of science but useless in classifying the sentences behind the lines. It is a trait of events and not of sentences."[94] Even if we could determine which sentences had originally been adopted as true by legislative postulation so that we might forever identify them as true by convention, this would be of no help, Quine explains. This would be mere historical conjecture, contributing little to our current epistemological understanding of our scientific theories: "Legislative postulation contributes truths which become integral to the corpus of truths; the artificiality of their origin does not linger as a localized quality, but suffuses the corpus."[95] Once such legislative postulates have been fully integrated into our theory, singling them out at some later date would be nothing more than discursive postulation.

Still, it might seem as though there is a difference with set theory in that its legislative postulation never fades, and thus set theory remains

[94] Ibid., p. 119.
[95] Ibid., pp. 119–20.

true by linguistic convention in a way that other sciences do not. However, Quine is very well aware of this point. Indeed, he considers set theory because of its very explicit and continuing appeals to convention. This does not run contrary to any of what he says about convention being a passing trait. Rather, it illuminates how exactly Quine does view that state of set theory as a science and why he may appear to be much more openly tolerant on this issue than, say, in his considerations of elementary logic. What distinguishes set theory for Quine is not so much its appeals to convention alone but rather what these appeals indicate: that set theory is at the moving front of science. He takes set theory to be a largely unsettled science and as such it remains caught up in legislative postulation.[96] This view would have been fairly uncontroversial from the discovery of the paradoxes through at least 1954. In coming to terms with the paradoxes, set theory had yet to return to a conception of set that was as natural as the original and, unfortunately, inconsistent one. Given this state, research in set theory proceeded in much in the way that Quine does it. It was experimental and exploratory; assorted hypotheses and axiom systems were put forward with an eye toward their various and conflicting consequences, and fundamental questions, such as the status of the axiom of choice and the continuum hypothesis, remained open. One of these axiomatizations might "become the common sense of another generation."[97]

Here, we might usefully contrast Quine's seemingly less tolerant attitude with regard to elementary logic, that is, classical first-order quantification theory.[98] Unlike set theory, elementary logic is largely a settled science, fully integrated into our current best theory. While Quine always remains

[96] Quine is explicit about this view of set theory in his 1962 "The Ways of Paradox" reprinted in *The Ways of Paradox*, remarking "that the discovery of antinomy is a crisis in the evolution of thought. In general set theory the crisis began sixty years ago and is not yet over" (p. 16). Although this view of set theory may have lessened later in his career, I think that this remained his basic attitude toward set theory throughout. I will return to this point in Chapters 6 and 7. This is also the sense in which Quine's notion of tolerance extends to all of science. At the moving front of science, that is, in areas that are largely unsettled, we put forward various hypotheses in hopes that one will eventually turn out to be best. But again, the positing of various hypotheses is for Quine part of the internal workings of science, which is to say that this is all part of scientific theorizing and not prior to it.

[97] Quine, "Whitehead and the Rise of Modern Logic," p. 27.

[98] While Quine always kept track of the differences between elementary logic – that is, first-order quantification theory – and set theory, he typically grouped them together under the single heading of logic, as he does in "Carnap and Logical Truth." I believe that his later insistence on not grouping set theory with logic is more a shift of emphasis that came with his growing focus on epistemological matters than a substantive change in his philosophical outlook, but I will not pursue this matter further here.

open to the possibility of revising elementary logic, at least in a legalistic sense,[99] he does not actively engage in developing new systems of logic, very much in contrast with his work in set theory. His grounds for this are the same that he applies in all areas of science: He thinks that no other purported logic brings with it, the familiarity, elegance, simplicity, and clarity of classical first-order quantification theory.[100] Contrary to Carnap, there is no need to be tolerant with regard to logic. From Quine's perspective, to continue to pursue alternatives as genuine logic would itself be an unscientific endeavor, much like continuing to pursue astronomy from the perspective of a geocentric universe. We have accepted classical first-order quantification theory as part of our best current theory and so count it as true. There is no external question here, only the ongoing progress that is internal to science as a whole.[101] Set theory, by contrast, was not nearly so settled a matter.

We might grant that Quine's position was reasonable – and even widely accepted – in 1954, but set theory has now matured, moving past the point at which it was at the moving front of science and becoming a much more settled matter. Quine himself was very much aware of this possibility, remarking that "Set theory, currently so caught up in legislative postulation, may some day gain a norm – even a strain of obviousness, perhaps– and lose all trace of the conventions in its history."[102] Perhaps set theory as embodied in the axioms of Zermelo-Fraenkel set theory with the axiom of choice (ZFC) has now reached this stage, making Quine's position obsolete, dogmatic, and unscientific. We could argue that, but we should note that set theory still faces many open questions. It was a great achievement to show that both the axiom of choice and the continuum hypothesis are independent of the Zermelo-Fraenkel (ZF) axioms. However, we might still want a set theory that settles these

[99] For a fairly late statement of this view, see Quine's "Reply to Jules Vuillemen," in *The Philosophy of W. V. Quine*, eds. Hahn and Schilpp, expanded edn, pp. 619–20.

[100] For an example of this attitude, see W. V. Quine, *Philosophy of Logic*, 2nd edn (Cambridge, MA: Harvard University Press, 1986), p. 87.

[101] Quine was well aware that there may have been reasons to reconsider logic in light of quantum mechanics, but on the whole, it seems that such a drastic revision has not maintained much of a hold on the philosophical imagination. Quine also has some sympathy with the constructive methods of the intuitionists (*Philosophy of Logic*, p. 88). It is surely an interesting mathematical endeavor to see what might still be proved by using only their more restrictive means, but this is not enough to recommend it as a wholesale replacement for our familiar logic. I take it that Quine would also not object to the exploration of alternative "logics" as abstract algebras, as many theories of pure mathematics are pursued with no immediate hope of application outside of themselves, whether in other areas of mathematics or in the natural sciences.

[102] Quine, "Carnap and Logical Truth," p. 120.

questions one way or another.[103] Furthermore, Quine's particular interest in the "big sets" also escapes investigation by way of ZF.[104] Finally, there are the non-well-founded sets, such as those that allow for self-membership, which are also completely left out of ZF, at least when the axiom of foundation is taken as one of its axioms, as it usually is.[105] I will address all of these topics in Chapters 6 and 7, which I hope will show that Quine's particular approach to set theory as explication is still very relevant and far from an unscientific dogma.

103 Quine remarks in a fairly late essay that he could see choice as being plausibly accepted as true but that he has very little in the way of intuitions for or against the continuum hypothesis. On this point, see his "Kurt Gödel," reprinted in his *Theories and Things*, pp. 145–6.

104 There was long an interest in such sets in the history of set theory – or at least in how we might talk of collections of absolutely everything in light of the thought that such collections lead to contradictions; on this point, see Hallett, *Cantorian Set Theory and Limitation of Size*, pp. 286–97. That they do not feature in much mainstream research in set theory seems to be the result of successful propaganda in favor of ZF rather than for compelling mathematical reasons. In fact, these sets have been found to feature in interesting ways in mathematics. See, for example, Saunders MacClane, "Locally Small Categories and the Foundations of Set Theory," in *Infinitistic Methods: Proceedings of the Symposium on Foundations of Mathematics* (New York, NY: Pergamon Press, 1961), pp. 25–43; and Saunders MacClane, "Categorical Algebra and Set-Theoretic Foundations," *Proceedings of the Symposium of Pure Mathematics*, AMS XIII, Pt. I, pp. 231–40.

105 Here, too, interesting and important mathematical research has emerged in recent years.

PART III

New Foundations and the Philosophy of Set Theory

CHAPTER 6

The Iterative Conception and Set Theory

The previous five chapters were largely concerned with historical and expositional issues related to set theory, in particular with the set theory and philosophy of W. V. Quine. In this chapter, I will continue with such themes, but here I intend to bring these previous considerations to bear on some more general issues in the philosophy of set theory. In particular, I will argue against the iterative conception of set, so often taken as exemplified by Zermelo-Fraenkel set theory (ZF), as the single correct version of set theory. Given the variety of set theories discussed in previous chapters, it might seem surprising that there should be a single correct version of set. Since the late 1960s, however, set theorists have tended to give the iterative conception of set, as expressed by ZF, privileged status, as if it captures something of the essence of *set*, if you will. There are various reasons why ZF has gained this status; perhaps foremost among them has been George Boolos's excellent exposition of the iterative conception in his 1971 essay "The Iterative Conception of Set."[1] In this paper, Boolos not only makes this conception accessible to philosophers and mathematicians alike but also does it in such a compelling manner that it is hard not to believe that there is something truly significant about the notion of set in the iterative conception.[2]

However, I will deny that the iterative conception is the only viable notion of set. In fact, this recent favoritism for a single conception of set

[1] George Boolos, "The Iterative Conception of Set," in *Logic, Logic, and Logic*, ed. Richard Jeffrey, with Introductions and Afterward by John P. Burgess (Cambridge, MA: Harvard University Press, 1998), pp. 13–29.

[2] It seems to me that there is something of the last vestige of the a priori in treating ZF in this way. In a sense, it allows philosophers to hold onto the idea that mathematics and its truths are of a unique and special variety among all the sciences, not subject to the sort of pragmatic theory-building constructions that philosophers seem more willing to admit for the traditionally empirical sciences. Indeed, Hao Wang talks of the intrinsic necessity of set theory depending on the iterative model; see Hao Wang, "The Concept of Set," in *Philosophy of Mathematics: Selected Readings*, eds. Paul Benacerraf and Hilary Putnam, 2nd edn, (New York, NY: Cambridge University Press, 1983), p. 553.

is, as perhaps already suspected, the anomaly in the history of set theory. As we have seen, the development of set theory after the paradoxes as well as before them has largely been a matter of competing intuitions about sets. Recall that Georg Cantor, though operating only with an informal conception of set, perhaps escaped the paradoxes because his prior theological views barred him from allowing what he deemed absolutely infinite sets, among these the universal set, the set of ordinal numbers, and the set of cardinal numbers. Bertrand Russell, by contrast, struggled to find a consistent notion of set in part because his own intuitions led him so naturally to the idea of a universal set. Cantor's religious views certainly held no sway for Russell, but he could initially find no principled way to rule out the problematic sets. Finally, Ernst Zermelo accepted Cantor's restrictions but, as we have seen, on largely pragmatic mathematical grounds. Like many logicians working in this early period of set theory, Zermelo could see that this was a mathematically interesting theory, and as he saw it, the aim of axiomatization was merely to capture enough of this theory to ensure its continued interest for mathematical research but to rule out enough that the theory would not give rise to contradiction. As we have seen, Russell, too, would ultimately agree, at least to an extent, that this pragmatic criterion was what would guide the further development of set theory.

In this chapter and Chapter 7, I will argue that such pragmatic concerns associated with the view that I have identified as set theory as explication should continue to be the primary factor in developing set theory. Indeed, I will argue that such pragmatic motivations are much more in line with both the historical and contemporary development of set theory, and attempting to restrict set theory to ZFC potentially has the harmful outcome of inhibiting the growth of our mathematical and scientific knowledge.[3] Therefore set theories such as Quine's New Foundations (NF) cannot and should not be ruled out on the grounds that they somehow stray too far from what was originally intended in the notion of set. In fact, what I hope to have brought out already and will continue to bring out here is that the idea of a single intuitive notion of set, especially as the iterative notion, is largely a myth of set theory's founding. I do not want to deny that there is some body of theory that

[3] Joseph Ullian has emphasized that Quine's exploratory and experimental approach to set theory is one that encourages the growth of knowledge rather than stifling it. See his "Quine and the Field of Mathematical Logic," in *The Philosophy of W. V. Quine*, eds. Edwin Hahn and Paul A. Schilpp (Chicago, IL: Open Court, 1998), p. 586.

we can pick out as set theory. Categorization is certainly a useful practice.[4] But to the extent to which we can offer criteria for picking out what counts as set theory, I will argue that NF, too, meets such criteria and furthers our understanding of sets in general.

6.1 Boolos's Exposition

To begin this chapter, let us first turn to George Boolos as perhaps the most important proponent of the iterative conception of set, owing, in no small degree, to the clarity of his exposition of this conception.[5] Boolos begins his paper by examining Cantor's proposed definitions of a set. Recall that Cantor defined a set as "any collection into a whole ... of definite and separate objects ... of our intuition or thought," or as "every aggregate of determinate elements which can be united into a whole by some law."[6] While Boolos rightly observes that these gesturings at the definition of set are variously unclear, he also states,

> But it cannot be denied that Cantor's definitions could be used by a person to identify and gain some understanding of the sort of object of which

[4] As Quine himself remarked in "Necessary Truth" in *The Ways of Paradox and Other Essays*, rev. and enlarged edn (Cambridge, MA: Harvard University Press, 1976), p. 76, "Boundaries between disciplines are useful for deans and librarians, but let us not overestimate them – the boundaries. When we abstract from them, we see all of science – physics, biology, economics, mathematics, logic, and the rest – as single sprawling system, loosely connected in some portions but disconnected nowhere."

[5] Boolos gave voice to a view of sets that had been around in the literature but was not widely known among philosophers or given much prominence among set theorists; see Boolos, "The Iterative Conception of Set," p. 16, fn. 3. Joseph Shoenfield sketched the view before Boolos, though in less detail, in his *Mathematical Logic* (Natick, MA: A. K. Peters, 1967), pp. 238–9, and again later in his "Axioms of Set Theory," in *Handbook of Mathematical Logic*, ed. Jon Barwise (New York, NY: North-Holland, 1977), pp. 323–4. After Boolos, Dana Scott also gave a rather detailed account of the conception in his "Axiomatizing Set Theory," in *Axiomatic Set Theory*, vol. II, ed. Thomas Jech (Providence, RI: American Mathematical Society, 1974), pp. 207–14. Kurt Gödel gives some early expression to this idea already in his 1947 "What Is Cantor's Continuum Problem?" in *Philosophy of Mathematics: Selected Readings*, eds. Paul Benacerraf and Hilary Putnam, 2nd edn (New York, NY: Cambridge University Press, 1983), pp. 474–5. Here, Gödel distinguishes the iterative conception from the conception of sets as extensions of predicates, claiming that "the perfectly 'naïve' and uncritical working with this concept of set [the iterative] has so far proved entirely self-consistent" (pp. 474–5). This might be interestingly contrasted with the attitude he expressed just seven years early, when he claimed that the axiom of foundation, which says there are no infinitely descending epsilon chains and hence no self-membered sets, is assumed only on pragmatic grounds of simplifying the work at hand. See his "The Consistency of the Axiom of Choice and of the Generalized Continuum Hypothesis with the Axioms of Set Theory," in *Kurt Godel: Collected Works*, vol. II, eds. Solomon Feferman et al. (New York, NY: Oxford University Press, 1990), p. 38. As I stated in the introduction to this chapter, it was largely following Boolos's masterly exposition that this conception rose to the fore as *the* conception of set.

[6] Cantor, *Contributions*, p. 85; *Foundations*, p. 916, n. 1.

> Cantor wished to treat. Moreover, they do suggest – although it must be conceded, only very faintly – two important characteristics of sets: that a set is "determined" by its elements in the sense that sets with exactly the same elements are identical, and that, in a sense, the clarification of which is one of the principal objects of the theory whose rationale we shall give, the elements of a set are "prior to" it.[7]

To illustrate what Cantor might have meant by these hints about what sets are, Boolos first sketches out what has become known as naïve set theory. We have seen this conception of a set many times previously. This is the idea that sets are determined by predicates or are the extensions of predicates. Recall that any given monadic predicate is either true of an object or false of it. We call the collection of things of which the predicate is true its extension. It is in this sense that a set can be thought of as the extension of a predicate. The idea is very intuitive and, as Boolos notes, seems to make sense of Cantor's remark that sets are collections of definite elements united by some law. But as we saw in previous chapters, despite its intuitiveness, the naïve conception is also inconsistent, as shown by the predicate "is non-self membered," which leads directly to Russell's paradox. Thus the naïve conception of set must be given up as making sense of Cantor's proposed definitions of set.[8]

"Faced with the inconsistency of naïve set theory," Boolos suggests, then,

> one might come to believe that any decision to adopt a system of axioms about set would be *arbitrary* in that no explanation could be given why the particular system adopted had any greater claim to describe what we conceive sets and the membership relation to be like than some other system, perhaps incompatible with the one chosen. One might think that no answer could be given to the question: why adopt *this* particular system rather than that or this other one? One might suppose that any apparently consistent theory of sets would have to be unnatural in some way or fragmentary, and that, if consistent, its consistency would be due to certain provisions that were laid down for the express purpose of avoiding the paradoxes that show naïve set theory inconsistent, but that lack any independent motivation [Boolos's italics].[9]

I quote this passage at length because it illustrates especially well the emerging debate between proponents of the iterative conception and a more pluralistic and experimental understanding of set theory viewed that was developing in response to the paradoxes. The picture that

[7] Boolos, "The Iterative Conception of Set," p. 13.
[8] Ibid., p. 14–16.
[9] Ibid., p. 16.

Boolos sketches here can clearly be identified with certain views of Quine but I think also with Zermelo and Russell, even if perhaps to a lesser extent or less explicitly for the latter two. But despite certain claims from all of them on the apparent ad hoc nature of axiomatizing set theory in light of the paradoxes, I think that this is not entirely true. There is certainly a guiding intuition behind various approaches to set theory presented by Zermelo, Russell, and Quine – namely, they all aim to save as much as possible of the idea that a set is the extension of a concept. Indeed, Quine's approach can be seen as a generalization of what is common to both Zermelo's and Russell's set theories, but let us finish laying out Boolos's position before returning to these issues.

Boolos goes on to remark that this artificiality does not have to be the case, as there is another conception of set, the iterative conception, that strikes him as entirely natural.[10] In this conception, much as in Russell's type theory, the sets are formed at stages, or levels. Unlike in Russell's theory, the stages are cumulative. A set will be any collection that is formed at a stage. We begin with individuals – again, as Russell did – and at stage zero, we form all collections of individuals. If no individuals exist, we form only the empty set at this stage. This will give us all subsets of individuals, including, of course, the set of all individuals itself. So we get 2^n sets at this stage, where n is the number of individuals and may be any natural number or even infinite. We now repeat this process at stage one, again for any individuals there may be and for all sets that are formed at stage zero. This process continues for all stages through that natural numbers. This brings us to the stage following stages 1, 2, 3, … , that is, stage omega, where we form all sets of sets of the previous levels, including a set of all sets formed at these earlier stages. We then continue as before, forming sets in this way through stages omega plus 1, omega plus 2, and so on to omega plus omega (or omega times 2), then to omega plus omega plus omega (omega times 3), and so on to omega times omega, and so on. In this system, the sets are reformed at each stage later than the stage that it was originally formed at. To simplify matters, we say that a set is formed only once, at the stage at which it was originally formed. Boolos also remarks that ZF generally does not assume individuals, so the sets over which it quantifies are the pure sets – the sets formed in the absence of any individuals – thus beginning with the empty set ∅, then forming {∅}, then {∅, {∅}}, and so on through the stages as before.[11]

[10] Ibid., p. 16.
[11] Ibid., pp. 18–20.

Boolos next axiomatizes this stage theory with the aim of ultimately showing how the axioms of ZF follow from it. He presents the following nine axioms with variables "x," "y," "z," … ranging over sets; "r," "s," "t," ranging over stages; the predicates of "$=$" and "$\in$" for equality and membership, respectively, as well as "E" for "is earlier than" and "F" for "is formed at":

(I) $(\forall s)\neg sEs$, no stage is earlier than itself;
(II) $(\forall r)(\forall s)(\forall t)((rEs\,.\,sEt) \rightarrow rEt)$, the transitivity of "earlier than";
(III) $(\forall s)(\forall t)(sEt \vee s = t \vee tEs)$, for any two stages, either one comes before the other or they are equal;
(IV) $(\exists s)(\forall t)(t \neq s \rightarrow sEt)$, there is an earliest stage;
(V) $(\forall s)(\exists t)(sEt\,.\,(\forall r)(rEt \rightarrow (rEs \vee r = s)))$, each stage is followed immediately by a stage;
(VI) $(\exists s)((\exists t)tEs\,.\,(\forall t)(tEs \rightarrow (\exists r)(tEr\,.\,rEs)))$, there is a stage aside from the first stage that does not immediately follow any stage, such as the omega stage;
(VII) $(\forall x)(\exists s)(xFs\,.\,(\forall t)(xFt \rightarrow t = s))$, every set is formed at a unique stage;
(VIII) $(\forall x)(\forall y)(\forall s)(\forall t)((y \in x\,.\,xFs\,.\,yFt) \rightarrow tEs)$, members of sets are formed before the set itself; and
(IX) $(\forall x)(\forall x)(\forall t)(xFx\,.\,tEs \rightarrow (\exists y)(\exists r)(y \in x\,.\,yFr\,.\,(t = r \vee tEr)$, if a set is formed at a stage, then at or after any earlier stage, at least one of its members has also been formed.

The first five axioms here are meant to govern the stages, while the last four describe at which stages sets and their members are formed. Boolos adds two additional axiom schemata: one for set specification and one for induction with regard to sets and stages. The specification axioms are of the form

$(\forall s)(\exists y)(\forall x)(x \in y \leftrightarrow (\chi\,.\,(\exists t)(tEs\,.\,xFt)))$, where χ is a formula of our language with no occurrence of "y" free.

These axioms state that for any stage, there is a set formed whose members are just those sets that were formed at earlier stages of which the formula χ is true. The induction axioms are of the form

$(\forall s)(\forall t)(tEs \rightarrow (\forall x)(xFt \rightarrow \theta)) \rightarrow (\forall x)(xFs \rightarrow \chi)) \rightarrow (\forall s)(\forall x)(xFs \rightarrow \chi)$, where χ is a formula of our language containing no occurrences of "t" and θ is like χ except that it contains free occurrences of "t" wherever θ contained occurrences of "s."

Intuitively, these axioms say that if a stage is covered by a predicate provided that all earlier stages are covered by it, then all stages are covered by that predicate. Boolos defines "a stage being covered by a predicate" as "the predicate holding of all sets formed at that stage."[12] The point of all this axiomatizing is to show next that we can derive the usual axioms of Zermelo set theory (Z) from this account of the stage theory. In this sense, Boolos claims that the iterative conception is a natural account of sets on a par with the naïve conception.

Recall that the axioms of Z are the empty set axiom, pairing, union, power set, infinity, separation, and extensionality. Boolos also includes regularity, or foundation, among these axioms. Extensionality, he says, "has a special status."[13] We will return shortly to each of these axioms in turn. Also, to simplify matters, as is usual, Boolos assumes that there are no individuals, only sets. We saw some version of the Zermelo axioms in Chapter 2, but to reiterate, Boolos states them as follows:

(1) Empty set: $(\exists y)(\forall x) \neg x \in y$, there is a set with no members;
(2) Pairing: $(\forall z)(\forall w)(\exists y)(\forall x)(x \in y \leftrightarrow (x = z \vee x = w))$, for any sets z and w, there is a set with z and w as its only members;
(3) Union: $(\forall z)(\exists y)(\forall x)(x \in y \leftrightarrow (\exists w)(x \in w \,.\, w \in z))$, for any set z, there is a set with just the members of z as its members;
(4) Power set: $(\forall z)(\exists y)(\forall x)(x \in y \leftrightarrow (\forall x)(w \in x \rightarrow w \in z))$, for any set z, there is a set with just the subsets of z as members;
(5) Infinity: $(\exists y)((\exists x)(x \in y \,.\, (\forall z) \neg z \in x) \,.\, (\forall x)(x \in y \rightarrow (\exists z)(z \in y \,.\, (\forall w)(w \in z \leftrightarrow (w \in x \vee w = x)))))$, calling z a successor of x if the members of z are only the members of x and x itself, there is a set containing an empty set and the successor of any set it contains;
(6) Separation: $(\forall z)(\exists y)(\forall x)(x \in y \leftrightarrow (x \in z \,.\, \varphi))$, where φ is a formula in which y does not occur free. For any set z, there is a set y of just those members of z for which φ holds;
(7) Regularity, or foundation: $(\exists x)\, \varphi \rightarrow (\exists x)(\varphi \,.\, (\forall y)(y \in x \rightarrow \neg \psi))$, where y does not occur free in φ and ψ is like φ except that it contains an occurrence of y wherever φ has a free occurrence of x. If there is a set x for which φ holds, then there is a set x for which φ holds but containing members y for which this formula does not hold.

Let us present just one example from Boolos of how the axioms of Z follow from the stage theory. Take, for example, the empty set axiom "$(\exists y)(\forall x)(\neg x \in y)$," which states that there is a set with no members.

[12] Ibid., pp. 20–22.
[13] Ibid., p. 25.

If we let χ be "$x = x$," this yields "$(\forall s)(\exists y)(\forall x)(x \in y \leftrightarrow (x = x \,.\, (\exists t)(tEs \,.\, xFt)))$," which is a specification axiom stating that for any stage, there is a set of all sets formed at some earlier stage. Since there is an earliest stage, there is a stage that has no sets formed before it. Hence there is a set with no members, so the empty set axiom follows from the stage theory. The other axioms of *Z* follow similarly, although replacement and choice do not.[14] Adding these two axioms gives us the full set theory ZFC, that is, Zermelo-Fraenkel set theory with the axiom of choice.

6.2 The Iterative Conception As Set Theory

There is no doubt something very appealing about the view of sets that we have just sketched, but in this section, I want to take a more critical attitude toward the iterative conception by looking at the reasons Boolos gives in its favor as well by trying to bring out some of its less intuitive features. Following his remarks on the supposed ad hoc nature of adopting any set theory in light of the paradoxes, Boolos remarks that the iterative conception "often strikes people as entirely natural, free from artificiality, not at all ad hoc, and one they might perhaps have formulated themselves."[15] This may be true, but so far, this only makes it a competitor with other conceptions of set. As we saw in Chapter 2, it might have seemed to Cantor entirely natural, for example, to rule out a universal set, given his background metaphysical views, but as we saw in that chapter, Russell found the opposite view so appealing that he could not see his way to fully accepting his own early proposal of type theory. Even with types in its initial *Principles of Mathematics* formulation, Russell thought that there must be a class of all terms, or, more generally, objects, and this would regenerate the paradoxes.[16] Similarly, Zermelo, for whom the set theory Boolos describes is named, put forward his axioms with a largely pragmatic justification aimed at preserving as much as was mathematically interesting from Cantor's original theory.[17]

[14] Ibid., pp. 22–6.

[15] Ibid., p. 16.

[16] Russell, *Principles*, pp. 366–7. A similar situation arises also for propositions (*Principles*, pp. 527–8). For more detail, see Hylton, *Russell, Idealism, and the Emergence of Analytic Philosophy*, pp. 229–31.

[17] Wang calls such approaches to restoring consistency to set theory "bankruptcy approaches" (see "The Concept of Set," pp. 541–4). He contrasts this with misunderstanding approaches, which search for some misunderstanding in our original understanding of a set. He claims Cantor as part of this tradition. I am not so sure. It seems to me that Cantor might never have understood set theory to embody something paradoxical. It is unclear whether he introduced his distinction between consistent and inconsistent totalities in reaction to paradox or whether this distinction was always in place, at least implicitly.

My point here is just to signal that intuitions alone, even the intuitions of set theory's founders themselves, will not settle questions about which set theory we are to accept. Boolos, I think, ultimately accepts this. Indeed, he continues, observing of the iterative conception,

> It is, perhaps, no more natural a conception than the naïve conception, and certainly not quite so simple to describe. On the other hand, it is, as far as we know, consistent: not only are the sets whose existence would lead to contradiction not assumed to exist in the axioms of the theories that express the iterative conception, but the more than fifty years experience that practicing set theorists have had with this conception have yielded a good understanding of what can and what cannot be proved in these theories, and at present there just is no suspicion at all that they are inconsistent.[18]

So in addition to its naturalness, the iterative conception appears to be consistent. But this seems to hold for any serious alternatives to the iterative conception. Certainly, type theory also appears to be consistent,[19] as does NF.[20] And while perhaps neither theory has received the scrutiny of the iterative conception as embodied in ZF, there does exist a vast amount of research on each, similar to the sort of research that set theorists have carried out on ZF. So consistency alone does not uniquely pick out the iterative conception from among other set theories, particularly in the experimental form that Boolos suggests here: that we think ZF consistent because in working with it, we have yet to find a contradiction. We should also flag another theme here that will continue to arise. Boolos remarks that "the sets whose existence would lead to contradiction [are] not assumed to exist in the axioms of the theories that express the iterative conception," but this raises another question. Are other sets, perhaps of either mathematical or philosophical interest, banned from these theories? The obvious possibility is that that none of them yield a universal set, in contrast to NF. This is an issue to which we will return shortly in more detail, considering NF more directly.

[18] Boolos, "The Iterative Conception of Set," p. 16.

[19] See, for example, Gerhard Gentzen, "The Consistency of the Simple Theory of Types," in *The Collected Papers of Gerhard Gentzen*, ed. M. E. Szabo (Amsterdam: North-Holland, 1969), pp. 214–22. Gentzen does not include infinity as one of the axioms of the simple theory of types, though he does include choice.

[20] We saw earlier Rosser's account that the usual sorts of contradictions do not arise in NF. Thomas Forster has conveyed to me in conversation that even a contradiction in NF would be exciting, since it would have to arise in wholly unexpected ways. Rosser, J. Barkley, "On the Consistency of Quine's New Foundations for Mathematical Logic." *The Journal of Symbolic Logic* (March 1939), pp. 15–24.

At this point, Boolos himself turns directly to NF (and its extension, ML). He remarks first that ZF is the standard first-order theory for expressing the iterative conception (also observing that the subsystem Z also embodies this idea as well as the extended systems Von Neumann-Bernays-Gödel (NBG) and Morse-Kelley (MK)). But he says, of theories that are proposed to be incompatible with ZF, by which he means NF and ML, that

> [t]hese theories appear to lack a motivation that is independent of the paradoxes in the following sense: they are not, as Russell has written, "such as even the cleverest logician would have thought of if he had not known of the contradictions." A final and satisfying resolution of the set-theoretical paradoxes cannot by embodied in a theory that blocks their derivation by artificial technical restrictions on the set of axioms that are imposed *only because* [Boolos's italics] paradox would otherwise ensue; these other theories survive only through such artificial devices. *ZF alone (together with its extensions and subsystems) is not only a consistent (apparently) but also an independently motivated theory of sets: there is so to speak, a "thought behind it" about the nature of sets which might have been put forth even if, impossibly, naïve set theory had been consistent* [my italics]. The thought, moreover, can be described in a rough, but informative way without first stating the theory the thought is behind.[21]

At least in part, Quine would accept Boolos's point, and so would Zermelo – and even Russell in certain moods. This was the point of Quine's remark that in light of the contradictions "commonsense is bankrupt." Any resolution of the paradoxes will have some degree of unnaturalness about it. But this is not to say that set theories such as type theory or NF are wholly unmotivated. The development of both types and NF is guided by capturing as much of comprehension in its original form as possible as the basis for set existence. ZF's separation is perhaps just that much farther from the original insight of set existence as a set being the extension of a predicate. Furthermore, type theory has very much the same sort of hierarchical structure that the iterative notion, as found in ZF, does, and NF takes its guiding thought from types and Russell's use of typical ambiguity in expositing his system. Indeed, we can even understand NF as taking this idea of a hierarchy, found in both type theory and ZF, and abstracting away from any actual layering of the universe. We can then view the insight of both types and ZF being not that the universe actually comes in layers but rather that

[21] Boolos, "The Iterative Conception of Set," p. 17. The Russell quote is from *My Philosophical Development* , p. 61.

the subscripting of variables can provide a syntactic test for set existence without the extra, perhaps metaphysical, claim that the universe comes in a prearranged hierarchy.[22] Both the idea of sets as the extensions of predicates and the ideas of sets being layered in a hierarchy point to ZF as not being the only independently motivated set theory with an easily describable thought behind it.

But Boolos thinks that the iterative conception brings out a very particular motivating thought that NF does not have. He next explains,

> Whatever tenuous hold on the conceptions of *set* and *member* were given one by Cantor's definitions of "set" and one's ordinary understanding of "element," "set," "collection," etc. is altogether lost if one is to suppose that some sets are members of themselves. The idea is paradoxical not in the sense that it is contradictory to suppose that some set is a member of itself, for, after all, "$(\exists x)(Sx \,.\, x \in x)$" is obviously consistent, but that if one understands "$\in$" as meaning "is a member of," it is very, very peculiar to suppose it true. For when one is told that a set is a collection into a whole of definite elements of our thought, one thinks: Here are some things. *Now* we bind them up into a whole. We don't suppose that what we come up with after combining some elements into a whole could have been one of the very things we combined (not, at least, if we are combining two or more elements) [Boolos's italics].[23]

NF certainly does allow for self-membered sets, for example, in its universal set V given by the instance of the comprehension schema "$(\exists y)(\forall x)(x \in y \leftrightarrow x = x)$." Then V is certainly a member of V because it is self-identical. This may very well seem counterintuitive in thinking about sets along the lines sketched by Boolos. As he presents it here, set theory sounds like a theory of collections of very ordinary physical objects: Here are some things, and we collect them together.[24] This sounds fine when we think

[22] In 1905, Russell also proposed the zigzag theory as a possible solution to the paradoxes stating, "In the zigzag theory, we start from the suggestion that propositional functions determine classes when they are fairly simple, and only fail to do so when they are complicated and recondite." However, he could not find criteria for determining when a propositional function became too complicated. See his "On Some Difficulties in the Theory of Transfinite Numbers and Order Types," pp. 145–51. Michael Potter observes that Quine's NF perhaps makes sense of Russell's zigzag theory and also observes this in his *Set Theory and Its Philosophy* (New York, NY: Oxford University Press, 2004), p. 54.

[23] Boolos, "The Iterative Conception of Set," pp. 17–18; Boolos's italics.

[24] Wang also talks this way in discussing the collections that make up set theory. He uses the example of two tables in a room and the idea that by looking at them, pointing at them, or thinking about them in the right sort of way, we might view them separately or as a unity; see "The Concept of Set," p. 531. Charles Parsons, in an attempt to get away from this idealistic view of how sets come to be, instead puts forth an account in terms of potentiality and actuality in his "What Is the Iterative Conception of Set?" in *Philosophy of Mathematics*, pp. 503–29. While Parsons's account is perhaps more in line with the Platonist tendencies of most set theorists, it depends upon our making rigorous sense of the controversial modal notions.

about, say, rocks or paper clips, but this brings us back to an issue that we have faced before: How intuitive is set theory? In its most intuitive version, it was, of course, inconsistent. But in the gloss that Boolos gives it here, it would hardly make sense of abstract collections of numbers or, say, of sets themselves: Here are some numbers. Now we bind them up into a whole. This is only a metaphor – perhaps a useful one, but a metaphor does not make rigorous sense of the concept of set theory as a mathematical theory of collections. I am at a loss, as I think we should be, when confronted by this situation along the same lines of our rock or paper clip collections. However, the abstract theory as embodied in the idea of a set as the extension of predicate makes reasonably clear the sense in which sets are collections of objects. This idea also seems to make sense of Cantor's remark about combining into a whole by a law. I can collect rocks, paper clips, numbers, and even sets according to whether or not a predicate is true of them. In its most general form, this idea led to contradiction. So intuitions are to be used with great caution. But this idea does guide us as to how set theory might use the notions of set and membership and the sense in which a set can be a member of itself. What it means for an object to be a member of a set is just for a certain predicate to be true of that object. This might not be our most ordinary understanding of set and membership, but it *is* one of the guiding thoughts that set theory began with.[25]

Boolos's metaphor seems to fail in other ways as well. What made set theory so important was its ability to make sense of the infinite. This is one of the minimal requirements of what any consistent set theory should do. How would our collecting together an infinite number of objects be made sense of according to our ability to bind objects together? Certainly, we cannot do this by running around grabbing up an infinite number of objects. Perhaps the metaphor helps if we think about a finite number of objects and then we can always just keeping adding one more. (I am not sure that it does help, especially as the idea that follows is that of the potential infinite, not the actual infinite as Cantor intended.) Even if this does seem to get us farther along in understanding infinite collections, it seems to fail again when we start talking about uncountable infinities and infinities of ever increasing size. None of this is meant to be definitive against Boolos, but it is to urge strong caution in our appeal to intuitions and metaphors when discussing set theory. The extent to which we have

[25] I am not suggesting that the extension of a predicate view is the only way in which we might make sense of set theory. I only urge it as a contender, and one that perhaps fares better when it comes to making sense of collections of objects, abstract or otherwise, in the set-theoretic sense or in the sense of self-membership.

any reliable intuitions about set theory seems to me very limited, and it can hardly be a deciding factor in determining which set theories are worthy of our investigations as areas of research. Indeed, much of what we find interesting about set theory runs quite contrary to our commonsense view of the world until we begin learning set theory.

Both of these issues raise a further question about the relationship of time to the iterative conception of set.[26] According to the iterative conception, as Boolos describes it, sets are formed at stages. So we have not only the question discussed earlier of how we collect objects into sets but also the question of when. It seems counterintuitive to think of sets being created at particular points in time, either past or future. Set theorists tend to work on the assumption that all the sets are already available to them, that they do not need to wait for some perhaps very large sets to be created at a future time (and since we can always "keep going" in the iterative hierarchy, it would seem that there are always sets not yet formed). Nor do they wonder at which point in the past a particular set came into being. Both questions sound absurd (although the question of when a particular set was discovered or the idea that more sets might be discovered in the future are not similarly absurd topics). Wang tries to put such questions aside by claiming that set formation, as the iterative conception describes it, is an idealization of human capacities, in terms of both our collecting abilities and time.[27] But we have to ask again: How ideal is this idealization? Again, it seems to rest on intuitions of which set theory was supposed to make sense in the first place. As Parsons observes, the kinds of intuitions that Wang would need to make sense of the iterative conception's idea of set formation would go not only beyond an intuition of the potential infinite – our ability to always perform a further step – but beyond limitations imposed by space–time structure itself.[28]

But perhaps Boolos admits that a battle among intuitions will not be resolved. As he concludes,

> There does not seem to be any argument that is guaranteed to persuade someone who really does not see the peculiarity of a set's belonging to itself, or to one of its members, etc., that these states of affairs are peculiar.

[26] Parsons, among others, has focused on this difficulty for the iterative conception. Trying to cope with this issue led him to his account of set existence in terms of potentiality and actuality. See his "What Is the Iterative Conception of Set?"

[27] Wang, "The Concept of Set," pp. 531–2.

[28] Parsons, "What Is the Iterative Conception?" p. 509. NF seems to fare better on the issues of set formation raised in this paragraph and on previous pages, since it allows us to say that the sets are already there. NF's comprehension principle then allows us to specify certain sets, but there is no talk of the sets being created by us or anything else.

> But it is in part the sense of their oddity that has led set-theorists to favor conceptions of set such as the iterative conception, according to which what they find odd does not occur.[29]

Again, we might say that "oddity" is a relative term, and much of what *any* set theorist would say to the uninitiated would sound quite odd indeed!

6.3 The Iterative Conception and Zermelo's Axioms

I want now to consider more carefully some of the axioms of Z that both follow and do not follow from the stage theory. I will begin with replacement and extensionality, as neither follows from the iterative conception and this seems quite uncontroversial. However, both replacement and extensionality raise interesting questions about whether the iterative conception deserves the privileged status it has gained. I will then turn to regularity, or foundation, which Boolos says does follow from the stage theory. This will bring us back to one of the issues that I considered in Section 6.2: that it is too strange to say that a collection can be a member of itself. The axiom of choice will be the last of the axioms I wish to examine. It, too, does not follow from the iterative conception, but as this leads to much broader issues concerning NF and set theory more generally, I will postpone discussion of the axiom of choice until Chapter 7.

Let us begin with replacement, since its absence from the iterative conception will be the least controversial. ZF, as Boolos describes it, consists of the previous axioms of Z plus the replacement axiom schema:

F is a function $\rightarrow (\forall z)(\exists y)(\forall x)(x \in y \leftrightarrow (\exists w)(w \in z \,.\, F(w) = x))$, if F is a function, then if its domain is a set, so is its image.

Boolos thinks it possible that we could have allowed an extension of the stage theory from which replacement would follow such as all instances of the following:

> If each set is correlated in some way with at least one stage, then for any set z there is a stage s such that for each member w of z, s is later than some stage with which w is correlated.

But while Boolos thinks that this "is an attractive further thought about the interrelation of sets and stages … it does seem to us to be a *further* thought, and not one that can be said to have been meant in the rough

[29] Boolos, "The Iterative Conception of Set," p. 18.

description of the iterative conception" [Boolos's italics].[30] For example, it could be that there are exactly ω_1 stages, since nothing in the rough account of the stage theory rules this out. In such a case, replacement would not hold generally. On the other hand, replacement has some well-known desirable consequences for the theory of sets. For example, it allows us to define a sequence of sets $\{R_\alpha\}$ with which each stage can be identified if we let $R_0 = \varnothing$; $R_{\alpha+1} = R_\alpha \cup P(R_\alpha)$ (the power set of R_α); and $R\lambda = \cup_{\beta>\lambda} R_\beta$, where λ is a limit ordinal. Replacement assures us that R is well defined, s is a stage if $(\exists\alpha)(s = R_\alpha)$, x is formed at s if $x \subseteq s$ and $x \notin s$, and that s is earlier than t if $(\exists\alpha)(\exists\beta)(s = R_\alpha$ and $t = R_\beta$, and $\alpha < \beta)$. Replacement allows us to go beyond translating the axioms of the stage theory into the language of set theory to stronger axioms asserting the existence of stages farther out than those suggested by the rough account of the stage theory.[31]

Echoing Russell's inductive justification of reducibility, Boolos then states,

> Although they are not derived from the iterative conception, the reason for adopting the axioms of replacement is quite simple: they have many desirable consequences and (apparently) no undesirable ones. In addition to theorems about the iterative conception, the consequences of replacement include a satisfactory if not ideal [fn. An ideal theory would decide the continuum hypothesis, at least.] theory of infinite numbers, and a highly desirable result that justifies inductive definitions on well-founded relations.[32]

While this remark is perfectly acceptable for a view of set theory carried out in a pragmatic and experimental spirit, it seems to run somewhat contrary to Boolos's declared preference for the iterative conception over all other versions of set theory.[33] While the iterative conception is perhaps intuitive and easy to describe, we see here that certain important mathematical features do not follow from it until we add the replacement axiom.[34] Again, adding the axiom seems perfectly acceptable if our guiding thought in developing set theory is to maintain as much as possible of what is mathematically interesting about it in light of the paradoxes. But if set theory is supposed to be

[30] Ibid., pp. 26–7.

[31] Ibid., p. 27.

[32] Ibid. We will return shortly this idea of an ideal theory in considering the axiom of choice.

[33] Of course, stronger claims have made for favoring certain axioms of set theory. Gödel and Wang both talk of axioms "forcing themselves upon us," although it does not seem entirely clear what this forcing consists of. See Kurt Gödel, "What Is Cantor's Continuum Problem?," p. 484, and Wang, "The Concept of Set," p. 552, respectively.

[34] In particular, the iterative conception, as Boolos describes it so far, without replacement leaves us with a set theory in which some sets do not have a cardinal number of members. For details, see W. D. Hart, *The Evolution of Logic* (New York, NY: Cambridge University Press, 2010), pp. 75–6. Such a set theory, to use Hart's words, "would be crippled."

grounded in the iterative conception to the exclusion of all other conceptions of set theory, we seem to be caught at an uncomfortable halfway point. On the one hand, we privilege ZF over all other set theories because it follows from the iterative conception and so gives expression to this intuitive notion of set formation. On the other hand, we see that the iterative conception is not strong enough to capture important mathematical features of set theory. Thus we are left adding axioms in a somewhat ad hoc way so as to restore the power that was lost in restricting the conception of set to avoid contradiction. I do not mean to suggest that we should now instead privilege some other version of set theory – say, NF –, over ZF. I simply want to point out that the intuitive picture that was supposed to lead us naturally to prefer ZF only gets us so far, mathematically speaking.

Let us next consider the axiom of regularity, or foundation, which Boolos says is among the axioms of Z that follow from the iterative conception. It most certainly does follow, as is demonstrated by Boolos's proof sketch. Let φ hold of a set x'. Then x' is formed at some stage that is not covered by $\neg\varphi$, since $\neg\varphi$ is false of x'. By an induction axiom, then, there is an earliest stage s not covered by $\neg\varphi$, though all stages earlier than s are covered by $\neg\varphi$. Hence there is an x formed at stage s of which $\neg\varphi$ does not hold and therefore of which φ does hold. If y is in x, then y is formed before stage s, so the stage at which y is formed is covered by $\neg\varphi$. Hence $\neg\varphi$ holds of y, and this is what $\neg\psi$ says.[35] What is tendentious about this is Boolos's claim that the regularity axiom is in fact an axiom of Z.

Recall from Chapter 2 that Zermelo does not assume this axiom as one of the axioms of his system. Indeed, it was not explicitly proposed as an axiom until von Neumann's work of the mid-1920s, and Zermelo assumes it on the purely pragmatic grounds of simplifying his set theory in light of the particular proof he is carrying out.[36] Zermelo himself was the first to explicitly adopt foundation as a general axiom of set theory but did not do so until 1930.[37] At least as late as 1973, Fraenkel, Bar-Hillel, and Levy state it merely as optional among axioms for Z.[38] So while the idea might look intuitive now in light of the iterative

[35] Boolos, "The Iterative Conception of Set," pp. 24–5.

[36] John von Neumann, "Ueber eine Widerspruchfreiheitsfrage in der axiomatischen Mengenlehre," in *John von Neuman: Collected Works*, vol. I, ed. Abraham H. Taub (New York, NY: Pergamon Press, 1961), pp. 494–508.

[37] Ernst Zermelo, "On Boundary Numbers and Domains of Sets: New Investigations in the Foundations of Set Theory," in *From Kant to Hilbert: A Source Book in the Foundations of Mathematics*, vol. II, ed. William Ewald (Oxford: Clarendon Press, 1996), pp. 1219–33.

[38] Fraenkel, Bar-Hillel, and Levy, *Foundations of Set Theory*, pp. 52–3. A little later in the work, they remark that since foundation "is not essential for mathematics, it cannot be regarded as fundamental by the traditional axiomatic attitude" (p. 89).

conception and in light of the paradoxes, it was, oddly enough, not an intuition about sets that occurred to many at the founding of the theory. Indeed, while it might not be a further thought about the iterative conception, it does seem a further thought about set theory in general.

What, then, recommends regularity? In short, the axiom says that there are no self-membered sets or sets with circular membership conditions, that is, no sets such that $x \in y$ and $y \in x$. Recall that Boolos argued that it is too strange to say that a collection can be a member of itself. But I countered that the metaphor driving this intuition does not hold up. Boolos offers no justification for it other than that it follows from the stage theory he sketches. I do not think that this is necessarily unfair in all cases as a justification, but here, it prejudices our choice of set theory. While other set theories also express much of what the ZF axioms do, a set theory that allows for non-well-founded sets would be automatically ruled out by regularity. If what it is to be a set is to be well founded, then a set theory that includes non-well-founded sets is ruled out from the start. Boolos's claim, then, about ZF alone being an apparently consistent and independently motivated account of sets seems to stand automatically unchallenged. Starting with the iterative conception, as Boolos describes it, drives us to accept ZF (including regularity) and its subsystems and extensions as *the* set theory.[39]

But there are other motivations for accepting regularity as an axiom of set theory. Although it seems that Mirimanoff first entertained the idea of focusing on well-founded sets in 1917, followed by von Neumann in his 1925 "An Axiomatization of Set Theory," it was Zermelo who fully accepted regularity as an axiom in his 1930 "On Boundary Numbers and Domains of Sets."[40] Neither von Neumann nor Zermelo claimed that the universe of sets was actually restricted to well-founded sets.

[39] For another sort of response to Boolos, see Thomas Forster, "The Iterative Conception of Set," *Review of Symbolic Logic*, 1:1 (2008), pp. 97–110. Forster argues here that the iterative conception is broader than cumulative hierarchy and includes some non-well-founded sets.

[40] Dimitri Mirimanoff, "Les Antinomies de Russell et de Burali-Forti et le Problème Fondamental de la Théorie des Ensembles," *L'Enseignement Mathématique*, 19 (1917), pp. 37–52, and his "Remarques sur la Théorie des Ensembles," *L'Enseignement Mathématique*, 19 (1917), pp. 208–17; John von Neumann, "An Axiomatization of Set Theory," in *From Frege to Gödel: A Source Book in Mathematical Logic, 1879–1931*, ed. Jean van Heijenoort (Cambridge, MA: Harvard University Press, 1967), pp. 393–413; Zermelo, "On Boundary Numbers and Domains of Sets," pp. 1219–33. Details of this history can be found in various locations; see, for example, Akihiro Kanamori, "The Mathematical Development of Set Theory from Cantor to Cohen," *The Bulletin of Symbolic Logic* 2:1 (1996), pp. 26–9; Potter, *Set Theory and Its Philosophy*, pp. 51–3; or Adam Rieger, "Paradox, ZF, and the Axiom of Foundation," in *Vintage Enthusiasms: Essays in Honor of J. L. Bell*, eds. Peter Clark, Michael Hallett and D. DeVidi (Glasgow ePrints Service: http://eprints.gla.ac.uk/3810, 2008), p. 10.

Rather, they both introduced the axiom for the practical purpose of obtaining categoricity results for their axiom systems. Regularity simplified the universe by eliminating non-well-founded sets, but none of them saw well-foundedness as essential to the concept of set. As Zermelo made the point with regard to accepting regularity: "This … axiom, which excludes all 'circular' sets, and all 'sets that contain themselves', and in general all 'groundless' sets, has always been satisfied in all practical applications of set-theory. Thus, for the time being, it presents no essential restriction to the theory."[41]

More recent justifications for regularity rest on similar grounds in that regularity simplifies the set-theoretic universe particularly with regard to inductive definitions and investigating models of set theory. This attitude is standard in set theory texts.[42] As is well known, all of the other axioms of ZF hold, regardless of regularity. Furthermore, all ordinary mathematics takes place within a universe of well-founded sets. So with regard to ease of use as well as perhaps Occam's Razor, regularity may seem quite worthwhile to assume, regardless of whether we think that this is inherent in the notion of set. Jech makes this point in his standard introduction to set theory:

> It should be stressed that, whether or not one accepts the Axiom of Foundation, makes no difference as far as the development of ordinary mathematics in set theory is concerned. Natural numbers, integers, real numbers and functions on them, and even cardinal and ordinal numbers have been defined, and their properties proved in this book, without any use of the Axiom of Foundation. As far as they are concerned, it does not make any difference whether or not there exist any non-well-founded sets. However, the Axiom of Foundation is very useful in investigations of models of set theory.[43]

Kanamori goes somewhat further in his history of set theory, remarking that regularity is in fact what makes set theory its own special branch of mathematics:

> It is nowadays almost banal that Foundation is the one axiom unnecessary for the recasting of mathematics in set-theoretic terms, but the axiom is also the salient feature that distinguishes investigations specific to set theory as an autonomous field of mathematics. Indeed, it can be fairly said

[41] Zermelo, "On Boundary Numbers and Domains of Sets," p. 1220.

[42] See, for example, Karel Hrbacek and Thomas Jech, *Introduction to Set Theory*, 3rd rev. and expanded edn (New York, NY: Marcel Dekker, 1999), p. 259, and Kenneth Kunen, *Set Theory: An Introduction to Independence Proofs* (New York, NY: North-Holland, 1980), pp. 100–1.

[43] Hrbacek and Jech, *Introduction to Set Theory*, p. 259.

> that current set theory is at base the study of well-foundedness, the Cantorian well-ordering doctrines adapted to the Zermelian generative conception of sets.

However, he is also careful to note that it is "a notable inversion [that] this iterative conception became a heuristic for motivating the axioms of set theory generally."[44]

So there do seem to be some distinctly mathematical reasons for preferring a set theory of well-founded sets as given by regularity. But is this attitude toward the well-founded sets an accurate picture of current set-theoretic research? It seems not. Particularly notable in this context is Aczel's work on non-well-founded sets.[45] Since its publication, interest in non-well-founded set theory has become widespread, with important applications in computer science. Philosophers, too, have recognized that such set theories have important applications, particularly in modeling ordinary language and in coming to grips with the semantic paradoxes.[46] Many of the current texts on set theory now include sections dedicated to non-well-founded sets.[47] Furthermore, Kanamori himself has noted that set theory itself has now become an autonomous subject of mathematical investigation[48] rather than merely serving as a foundation for all other mathematics. If we view set theory in this way, it seems that we should be interested in all sets, not just the well-founded ones. Should we not also want to broaden our investigations to understanding what the non-well-founded ones are like? So it seems that non-well-founded sets do have a place in the mathematics and philosophy surrounding set theory. Indeed, it is perhaps fair to say that out-of-hand dismissals of such set theories have impeded the acquisition of mathematical knowledge.

[44] Kanamori, "Mathematical Development of Set Theory," pp. 27–8.

[45] Peter Aczel, *Non-Well-Founded Sets* (Stanford, CA: CSLI Publications, 1988).

[46] The most notable work here is Jon Barwise and John Etchemendy, *The Liar: An Essay on Truth and Circularity* (New York, NY: Oxford University Press, 1987). W. D. Hart has suggested that we might use set theory with a universal set to give a better account of the universal quantifier, as this would give a sense in which "all" really does mean *all*. See his *The Evolution of Logic*, pp. 113–14. In a recent collection of essays edited by Agustin Rayo and Gabriel Uzquiano, *Absolute Generality* (New York, NY: Cambridge University Press, 2006), some mention is made of this, but the idea is quickly dismissed on grounds similar to what Boolos says at the beginning of his paper on the iterative conception. See in particular Oystein Linnebo, "Sets, Properties, and Unrestricted Quantification," in *Absolute Generality*, pp. 156–7; and Alan Weir, "Is It Too Much to Ask, to Ask for Everything?," in *Absolute Generality*, p. 340.

[47] See, for example, Hrbacek and Jech, *Introduction to Set Theory*; Yiannis Moschovakis, *Notes on Set Theory*, 2nd edn (New York, NY: Springer, 2006); and Devlin, *The Joy of Sets*. Rieger observes that the picture on the cover of Devlin's book is in fact of a non-well-founded set. See his "An Argument for Finsler-Aczel Set Theory," *Mind* 109:434 (2000), p. 241, fn. 1.

[48] Kanamori, "Mathematical Development of Set Theory," p. 1.

I want to turn now to the extensionality axiom and the special epistemological status that Boolos says that it has, which, like replacement, does not follow from the iterative conception. Boolos observes that if someone were to deny any other axiom of ZF, we would be more inclined to believe the axiom false than if someone were to deny extensionality. The claim "There are distinct sets with the same members" seems so far from any ordinary conception of set that it would justify our believing that the asserter of such a statement must have some non-standard usage of the word "set," that they are not merely claiming extensionality to be false. "Because of this difference," he concludes, "one might be tempted to call the axiom of extensionality 'analytic,' true by virtue of the meanings of the words contained in it, but not to consider the other axioms analytic."[49] Analyticity is, of course, a controversial notion, and Boolos recognizes this, noting that until we have an account of how a sentence can be true by virtue of meaning, we should refrain from classifying extensionality as analytic. Still,

> [i]t seems probable, nevertheless, that whatever justification for accepting the axiom of extensionality there may be, it is more likely to resemble the justification for accepting most of the classical examples of analytic sentences, such as "all bachelors are unmarried" or "siblings have siblings" than is the justification for accepting the other axioms of set theory. That the concepts of *set* and *being a member of* obey the axiom of extensionality is a far more central feature of our use of them than is the fact that they obey any other axiom. A theory that denied, or even failed to affirm, some of the other axioms of ZF might still be called a set theory, albeit a deviant or fragmentary one. But a theory that did not affirm that the objects with which it dealt were identical if they had the same members would only by charity be called a theory of *sets* alone [Boolos's italics].[50]

The important point here, regardless of analyticity, is that one feature we look for in identifying something as a set theory is that its objects should be identical when they have the same members. This seems correct to me. Much of the value gained by sets is that they have a very clear identity criterion in extensionality, and without this, it seems that we are dealing with some other kind of entity. But non-well-founded set theories do share this feature with the iterative conception. So if we pick out extensionality as characterizing sets, then why should we favor the iterative conception, especially as this key axiom does not follow from it? Indeed, we

[49] Boolos, "The Iterative Conception of Set," pp. 27–8.
[50] Ibid., p. 28.

could, according to Boolos, deny any of the other axioms and still end up with a fragmentary or deviant set theory.[51]

This does not seem entirely correct. Throughout the history of set theory, another axiom has also been cited as characterizing set theory: the comprehension axiom. This also does not follow generally from the iterative conception, but this is surely a virtue. In its most general formulation, it should not be part of any set theory, since unrestricted comprehension yields the set-theoretic paradoxes. However, the iterative conception does yield separation, ZF's restricted version of comprehension, and any set theory will need some such principle of set specification. Without it, we can not specify any sets. This boils set theory down to two identifying characteristics – extensionality and comprehension – and nothing about these two principles suggests the iterative conception. Indeed, they seem more likely to work against the iterative conception in the following way.

The naïve conception of set gave us too much in that it yielded contradictory sets, such as Russell's set of all non-self-membered sets. One fairly common way of judging a successful axiomatization is that it rules out these contradictory sets while doing as little damage as possible to the original theory. In this sense, the iterative conception as expressed by ZF perhaps rules out too much.[52] While we do not want contradictory

[51] There is a potential difficulty for non-well-founded sets and the extensionality axiom. Set theory with the regularity axiom assures us that there is a recursive algorithm for deciding identity between sets. While non-well-founded sets do not violate extensionality, this axiom must be somehow strengthened to account for identity between such sets. There are various ways in which this might be done, but it is not entirely clear that any one of them is to be preferred. Aczel raises this issue in his *Non-Well-Founded Sets* in considering the various anti-foundation axioms. For a concise exposition of the issue, see W. D. Hart, "On Non-Well-Founded Sets," *Crítica* XXIV:72 (1992), pp. 3–20. For a recent account of non-well-founded set identity in terms of games that is fairly analogous to the situation with well-founded sets, see Forster, "The Iterative Conception of Set," pp. 97–110. Quine himself recognized this issue at least as early as 1975 in his "On the Individuation of Attributes," in *Theories and Things* (Cambridge, MA: Harvard University Press, 1981), pp. 102–3. Here, he observes, "The system of my 'New Foundations' does have ungrounded classes, and so the system of my *Mathematical Logic*; and it could be argued that for such classes there is no satisfactory individuation. They are identical if their members are identical, and these are identical if their members are identical, and there is no stopping. This, then, is a point in favor of the systems that bar ungrounded classes" [Quine's italics].

[52] Rieger argues similarly in his "Paradox, ZF, and the Axiom of Foundation," p. 18. In a sense, perhaps something like this can be found already in Zermelo's "Investigations in the Foundations of Set Theory I" in his comment that in light of the paradoxes "there is at this point nothing left for us to do but to proceed in the opposite direction and, starting from set theory as it is historically given, to seek out the principles required for establishing the foundations of this mathematical discipline. In solving the problem we must, on the one hand, restrict these principles sufficiently to exclude all contradictions and, on the other, take them sufficiently wide to retain all that is valuable in this theory" (p. 200).

sets, we also had other sets in the original theory, such as the universal set specified by the instance of comprehension "$(\exists y)(\forall x)(x \in y \leftrightarrow x = x)$." ZF does away with this, as does type theory, at least in its most general version. (Recall that type theory does yield a series of quasi-universal sets.) However, NF does not while still apparently ruling out contradictory sets. Indeed, its only two axioms are versions of extensionality and comprehension. This is the same for type theory, but it has perhaps the undesirable complication of dividing the universe into levels, again perhaps moving beyond the most basic characteristics of sets. ZF's axioms seem perhaps even more ad hoc. Given the weakening of comprehension in the form of separation, many of the other axioms just serve as fixes for restoring the power that was lost from the original naïve theory. Starting with the iterative conception and moving to ZF does seem to provide some motivation for adopting ZF, but what I am trying to bring out here is that starting from our naïve conception of set and moving back to ZF makes ZF seems quite arbitrary and unmotivated as a set theory, at least with regard to capturing the original notion of set with which we began. I do not take any of these arguments to be definitive, but I do think that they show that any set theory can perhaps be made to look quite arbitrary, depending on what we take as our starting point for thinking about sets. We have seen this play out in a variety of ways, particularly in investigating the historical development of set theory.

This still leaves us to discuss the axiom of choice. I will postpone this until the Chapter 7, as it will lead us to much more general considerations of NF. For now, I will just sum up our considerations so far of Boolos's account of the iterative conception of set. The general point that I have tried to make throughout this chapter is that while other set theories might seem artificial, there is something quite artificial in ZF as well. We have seen this both in the assumption of replacement as an extra postulate and in ZF's ruling out certain seemingly legitimate and perhaps quite interesting sets, namely, the non-well-founded sets. Let us turn now to the axiom of choice, which will lead us to consider more generally NF and Quine's views on set theory as a whole.

CHAPTER 7

New Foundations, the Axiom of Choice, and Arithmetic

In this final chapter, we consider W. V. Quine's New Foundations (NF) more generally in comparison with other set theories, and we conclude with a return to his views on the philosophy of set theory as a whole. Recall from Chapter 6 that George Boolos claimed that ZF is the only apparently consistent and independently motivated set theory. I have been arguing against this claim, trying to point out the ways in which ZF is far less intuitive than Boolos presents it and the ways in which other set theories, including NF, are also independently motivated. Indeed, throughout this book, I have been suggesting that the development of set theory has been largely a battle of competing intuitions over what sets are like. This may be worrisome in that it perhaps leaves us with no objective way of determining what should count as a viable set theory. What we call set theory could just be a matter of personal preference. Should we allow just any theory of collections count as a set theory? However, the situation is not as bleak as that. There do seem to be some readily available criteria for assessing whether to count some body of theory as set theory.

We have already seen some of this in our recounting of the history of the development of set theory in that the concept of a set began with some idea of sets being the extensions of predicates (though initially, Georg Cantor says little about what sets are). While I take all set theories to have this origin – and the history seems to support this view – this in itself is perhaps a controversial claim. Many philosophers and logicians have stated that there are two separate notions of a set: the logical and the mathematical. The former takes sets to be the extensions of predicates, and the latter takes them to be formed according to the iterative conception.[1] Having touched upon this debate elsewhere, I will not say

[1] Kurt Gödel, for example, draws this distinction in his "What Is Cantor's Continuum Problem?" in *Philosophy of Mathematics, Selected Readings*, eds. Paul Benacerraf and Hilary Putnam, 2nd edn (New York, NY: Cambridge University Press, 1983), pp. 474–5.

more here other than to reiterate that I think that the distinction does not hold up very well and that it is mostly a distinction that encourages a privileging of the iterative conception. The usual argument is that the logical conception of set has its home in philosophy and that is where the paradoxes arose. Within mathematics, the iterative conception was always present, so mathematicians were never troubled by the paradoxes or the notion of a set more generally. However, the history shows the iterative conception and this distinction emerging only after the paradoxes (see Chapter 2 especially). I will now put this issue aside.

We saw in Chapter 6 that extensionality and comprehension give us some minimal account of what to count as a set. As we saw in previous chapters, I have suggested some additional criteria that any viable set theory should meet.[2] First, set theory must deal with the paradoxes in some way. Second, set theory should contain a mathematically interesting account of the infinite, as this was one of the aspects that drove the development of set theory in the first place. Therefore it should preserve at least some version of Cantor's theorem. Third, set theory should be capable of serving as a framework within which ordinary mathematics can be carried out. This seems a reasonable, minimal set of criteria for identifying a theory of collections as a viable set theory in the way the tradition intended set theory to be, although we might still revise this list in light of further development in set theory. We have already seen that NF succeeds on the first two criteria. The primary focus of this chapter will be to show that it succeeds also on the third. To begin this discussion, I pick up with Boolos's considerations of the axiom of choice in relation to the iterative conception.

7.1 The Axiom of Choice

Boolos introduces choice in the form often called "the multiplicative axiom,"[3] which states,

> For any *x*, if *x* is a set of nonempty disjoint sets, then there is a set, that is, a choice set for *x*, that contains exactly one member of each of the members of *x*.

[2] W. D. Hart suggested these to me in conversation, but Quine himself suggests something like this in his "Foundations of Mathematics," in *The Ways of Paradox and Other Essays*, rev. and enlarged edn (Cambridge, MA: Harvard University Press, 1976), pp. 31–2.

[3] W. D. Hart has pointed out to me that the name comes from Bertrand Russell, as he used this version of the axiom to show infinite products to be nonempty. For details, see Russell, *Introduction to Mathematical Philosophy* (London: George Allen and Unwin, 1919), p. 122.

The axiom is crucial for much of ordinary mathematics, but as Boolos rightly notes, neither choice nor its negation follows from the iterative conception. Much like replacement, he suggests, we might extend the stage theory so that it does decide choice (no doubt preferably that it be true). "But," he remarks,

> it seems that no additional axiom, which would decide choice, can be inferred from the rough description without the assumption of the axiom of choice itself, or of some equally uncertain principle, in the inference. The difficulty with the axiom of choice is that the decision whether to regard the rough description as implying a principle about sets and stages from which the axiom could be derived is as difficult a decision, because essentially the same decision, as the decision whether to accept the axiom.[4]

For example, we might try to extend the stage theory by adding the following principle:

> Let x be a set of nonempty disjoint sets, so x is formed at some stage s. Then the members of x must be formed at some stage earlier than s. Therefore at stage s or earlier, there exists a set that contains exactly one member of each member of x.

But such a principle, Boolos explains, begs the question, since we have no reason to think that such a set is in fact formed. It is the axiom of choice itself that allows us to choose exactly one member from each member of x, that is, to form a choice set for x.[5] He concludes, "To say this is not to say that the axiom of choice is not both obvious and indispensable. It is only to say that the justification for its acceptance is not to be found in the iterative conception of set."[6]

Again, as with the axiom of replacement, a crucial mathematical principle does not follow from the iterative conception of set. This is again not to say that we should reject ZF as a framework for mathematics; rather, we can get only so far motivating it by way of the iterative conception. The situation is the same for type theory; choice must be assumed as an additional axiom if the system is to be sufficient for ordinary mathematics. If we approach set theory in a pragmatic spirit, such additional postulates should be of minimal concern. Our guide to developing set theory is to incorporate enough power to make the system mathematically interesting. But the option of adding choice is not

[4] Boolos, "The Iterative Conception of Set," p. 28.
[5] Ibid., pp. 28–9.
[6] Ibid., p. 29.

available to NF. In fact, this will be our first serious challenge in adopting NF as a set theory, for NF does decide choice, and it says that this principle is false. So can NF serve as a plausible framework for mathematics? This is the question to which we now turn.

There is no doubt that the axiom of choice is an important principle for mathematics, used in a wide variety of results from all areas of the discipline.[7] But the axiom itself has had a rather controversial history. Ernst Zermelo, in his 1904 well-ordering paper,[8] was the first to put forward the principle as an explicit axiom, which he used to show that every set can be well ordered.[9] Attacks on the proof immediately followed, describing its use as an illegitimate method for mathematics. Zermelo responded with his 1908 proof and his fully developed axiom system for set theory.[10] Although choice finally won out as an accepted method of proof, its status remains somewhat different from that of other axioms of set theory. Justification for the axiom generally points to its usefulness and importance in mathematics. Indeed, much of Zermelo's strategy in defending it was to show that a significant amount of already accepted mathematics relies upon it, albeit as an implicit assumption. Such arguments continue to be common in set theory texts. For example, Jech points to important results that rely on choice such as the Hahn-Banach theorem, Tichonov's theorem, and the maximal ideal theorem as accounting for the axiom's "universal acceptance."[11] Despite this widespread acceptance, mathematicians and logicians continue to be careful about the use of choice. In particular, they often distinguish proofs in ZF from those done in ZFC (ZF plus choice) and frequently try to reprove results that rely on choice with proofs that do not. Indeed, much of Jech's most notable work focuses on such results.[12] Quine himself, perhaps unsurprisingly, adopts this attitude in *Set Theory and Its Logic*, remarking,

> Intuitively the principle seems reasonable, even obvious. The trouble is just that it is not as simple and elementary a statement as could be desired as a starting point, and no way is known of deducing it strictly from

[7] For some examples of the mathematics relying on choice, see Hrbacek and Jech, *Introduction to Set Theory*, pp. 144–53.

[8] Ernst Zermelo, "Proof That Every Set Can Be Well-Ordered," pp. 139–41.

[9] For a complete history of the axiom and the controversies that followed it, see Gregory H. Moore, *Zermelo's Axiom of Choice: Its Origins, Development, and Influence* (New York, NY: Springer, 1982).

[10] Indeed, Moore argues that the controversy over choice was the primary motivation for Zermelo's axiomatization. As I stated in Chapter 2, I do not fully agree with this account.

[11] Hrbacek and Jech, *Introduction to Set Theory*, p. 153.

[12] Thomas J. Jech, *The Axiom of Choice* (Mineola, NY: Dover, 1973).

> anything simpler. For this reason it is common practice to distinguish between results that depend on the axiom of choice and those that do not. Accordingly I shall not adopt the principle as axiom or axiom schema, but shall merely use it as premiss where needed.[13]

Aside from such justifications by way of its usefulness, there may also be something that seems rather obvious about the axiom. From a collection of disjoint sets, it allows us to pick a unique representative for each one, yielding a choice set for the collection. Jech observes that the axiom is often described by using the analogy of an election in which the sets in the collection are the candidates for a particular office and the election process gives us a choice function from each of these sets. But he is careful to point out that such an analogy talks only of a finite case and that this gives us no justification for assuming the axiom to hold in infinite cases.[14] Reflecting on the arithmetic of infinite numbers, for example, suggests in a pretty straightforward way that what holds for the finite should give us no reason to think that it will hold analogously for the infinite. Indeed, as I have pointed out several times, we seem to have very few initial correct intuitions about the infinite, and it was only Cantor's discovery of set theory that allowed us to gain an understanding of it.

It turns out that for the finite case, the axiom is unnecessary, since it is provable in dealing only with finite collections of finite sets. The situation is, of course, very different for infinite sets. Indeed, even when an infinite collection contains only finite sets, choice is not generally provable.[15] So choice must be added as an axiom to deal with infinite sets in general. Now we get some fairly unintuitive results. Stefan Banach and Alfred Tarski, generalizing a result of Felix Hausdorff, proved, using choice, that a sphere can be decomposed into parts and rearranged, using rigid motions, to obtain two spheres, each of which is the same size as the original sphere.[16] A popular gloss on the result is that it allows us to decompose a sphere the size of a pea and reassemble

[13] W. V. Quine, *Set Theory and Its Logic*. Rev. edn (Cambridge, MA: Harvard University Press, 1969), pp. 217–18.

[14] Thomas J. Jech, "About the Axiom of Choice," in *Handbook of Mathematical Logic*, ed. Jon Barwise (Amsterdam: North-Holland, 1977), p. 350.

[15] Ibid., pp. 350–1.

[16] Ibid., pp. 351–2. The original paper is Stefan Banach and Alfred Tarski, "Sur la Décomposition des Ensembles de Points en Parties Respectivement Congruents [1924]," in Alfred Tarski, *Collected Papers*, vol. I, 1921–34, eds. Steven R. Givant and Ralph N. McKenzie (Boston, MA: Birkhäuser, 1986), pp. 121–54. Many authors mention this result to illustrate that choice is not as intuitive as it may seem. There are other results of this kind, such as the existence of unmeasurable sets of real numbers, but such examples are not nearly as vivid for the nonspecialist.

it into a sphere the size of the sun.[17] This has become known as the Banach-Tarski paradox. The point here is not to claim that choice must be false in light of such a result, but merely to bring out its lack of grounding in common sense. As Michael Potter observes, the unintuitive nature of the result depends upon our trying to understand it in terms of our ordinary intuitions about elementary geometry. The argument for this paradoxical result does not rely on any such intuitions but rather depends on a mathematical understanding of transcendental functions.[18]

I have been continually coming back to such attempts at grounding set theory in ordinary everyday conceptions of, say, collections or of the infinite. Our understanding of such issues may develop in conjunction with certain initial intuitions about sets, but this involves a continual back and forth between the technical development of the subject and our initial intuitions about it. If intuitions alone were the final arbiter for the success of a mathematical theory, we would have to throw out much of mathematics as we now know it. Certainly set theory would have ended with Bertrand Russell's paradox. I also want to make the point that there is some unfairness in examining set theories other than ZF (or ZFC). Because they violate our ordinary understanding of, say, what it means to be a member of a collection, they are often treated with disdain and as being unworthy as research projects. I am suggesting that such nonstandard set theories are thought to be nonstandard, that is, contrary to intuition, only because we have become so familiar with the iterative conception as exemplified by ZF.

To return to our more immediate topic, what of the axiom of choice in NF? The situation does not initially look good, since Ernst Specker proved in 1953 that choice is refutable in NF.[19] We want to know exactly what Specker proved, though. Will it prohibit much of accepted mathematics? If so, this seems to be a serious point against NF as a framework within which mathematics can take place. If much of accepted mathematics can be developed in NF, however, this cannot be used as a point against NF. In fact, we might then have reason to prefer NF over

[17] For a modern technical account, see Stan Wagon, *The Banach-Tarski Paradox* (Cambridge: Cambridge University Press, 2005). This gloss can be found on p. xiii.

[18] Michael Potter, *Set Theory and Its Philosophy* (New York, NY: Oxford University Press, 2004), pp. 275–6.

[19] Ernst P. Specker, "The Axiom of Choice in Quine's New Foundations for Mathematical Logic," *Proceedings of the National Academy of Sciences*, 39:9 (September 1953), pp. 972–5.

ZF. Choice is independent of the axioms of ZF; we can neither prove it nor refute it in ZF. On this count, then, NF seems to be a more ideal theory, as Boolos characterizes an ideal theory, since it decides the axiom of choice, whereas ZF does not. If we were to accept NF as a working set theory, we would get both the development of accepted mathematics and an answer to the truth or falsity of choice. Let us see, then, what exactly the situation is.

Specker's refutation of choice in its original form is rather abstract and hard to grasp apart from what appear to be some rather arcane details of NF. To make the proof more accessible, I instead follow the account that J. Barkley Rosser gives in his review of Specker's result.[20] We will assume the axiom of choice and then proceed by reduction. Rosser begins by observing that the proof depends significantly on many properties of cardinals being invariant when their types are raised uniformly. That is, given a cardinal m, we choose a class α of cardinality m and let $T(m)$ be the cardinality of the singleton subclasses of α, USC(α). $T(m)$ thus gives us the cardinal m, but it is of one type higher. (We saw the importance of such type raising operations in NF in Chapter 3 in the discussion of Cantor's theorem.) To illustrate such invariance, Rosser gives a few examples:

$m \leq n \equiv T(m) \leq T(n)$;
$T(m+n) = T(m) + T(n)$, when $m+n$, m, and n are all cardinals; and
$T(0) = 0$, $T(1) = 1$, $T(2) = 2$, and so on.

As a consequence, we see that if $m = q + q + q + k$, where $k = 1$, 2, or 3, then $T(m) = T(q) + T(q) + T(q) + k$. So while residues (mod 3) are invariant, cardinality is in general not. In fact, we cannot even prove for all finite m that $T(m) = m$.

Next, to define 2^m so that it has the same type as m, we find an α, if such an α exists, such that the cardinality of USC(α) $= m$. We then let 2^m be the cardinality of all subclasses of α. Now if 2^m exists, then $m < 2^m$ and $T(2^m) = 2^{T(m)}$, another case of invariance. However, 2^m does not exist for all m. Indeed, if $|V|$ is the cardinality of the universe, then 2^m exists if and only if $m \leq T(|V|)$. We also note that $2^{T(|V|)} = |V|$, $2^{T(T(|V|))} = T(|V|)$, and so on. We also define, for any cardinal m, $\Phi(m)$ as the set (it is stratified) of m and as many of 2^m, 2^{2^m} ... as exist. If they all exist so that $\Phi(m)$ is infinite, then $\Phi(T(m))$ consists of $T(m)$, $T(2^m)$, $T(2^{2^m})$, ... and is also infinite.

[20] J. Barkley Rosser, "Review of Specker's 'The Axiom of Choice in Quine's New Foundations for Mathematical Logic,'" *Journal of Symbolic Logic*, 19:2 (1954), pp. 127–8.

The interesting case is when $\Phi(m)$ is finite. For example, $2^{|V|}$ does not exist, so $\Phi(|V|)$ contains only $|V|$.

By choice, we can show that if $\Phi(m)$ is finite, then $\Phi(T(m))$ is also finite and also that the cardinalities of $\Phi(m)$ and $\Phi(T(m))$ have different residues (mod 3). Let $\Phi(m)$ have finite cardinality. Then there is a greatest member of $\Phi(m)$, say, n. So 2^n cannot exist; otherwise, it would be the greatest member of $\Phi(m)$. Thus $T(|V|) < n$, since cardinal comparability follows from choice. Then $T(T(|V|)) < T(n)$, so $T(|V|) = 2^{T(T(|V|))} \leq 2^{T(n)}$. This gives us two cases to consider:

<u>Case 1</u>: $T(|V|) = 2^{T(n)}$. Then the members of $\Phi(T(m))$ are $T(m)$, $T(2m), \ldots, T(n), 2^{T(n)}, 2^{2^{T(n)}}$, so intuitively, $\Phi(T(m))$ has two more members than $\Phi(m)$. So $\Phi(T(m))$ is finite, and its residue (mod 3) differs from that of $\Phi(m)$.

<u>Case 2</u>: $T(|V|) < 2^{T(n)}$. Then $2^{T(n)}$ is the greatest member of $\Phi(T(m))$. So now $\Phi(T(m))$ has intuitively one more member than $\Phi(m)$, and, similar to the first case, $\Phi(T(m))$ is finite, and its residue (mod 3) differs from that of $\Phi(m)$.

By choice again, there is a least m for which $\Phi(m)$ is finite, and since $\Phi(\mathrm{T}(m))$ is also finite, $m \leq T(m)$. By the definition of $T(m)$, there is a p such that $m = T(p)$ and $p \leq m$. Hence $p = m$, so $T(p) = T(m)$, and then finally $m = T(m)$. Therefore the cardinals of $\Phi(m)$ and $\Phi(T(m))$ have the same residue (mod 3), which is a contradiction. Thus the axiom of choice is false in NF. Also, since the generalized continuum hypothesis implies choice, it too must be false.[21]

Where does this leave us with regard to NF as a framework for ordinary mathematics? It seems to leave us in a situation that is perhaps surprisingly similar to where we are with ZF. Rosser observes almost immediately in his note that he "has become increasingly convinced that" all ordinary mathematics can be developed within the Cantorian sets, that is, those sets α such that they and the set of their singleton subsets have the same cardinality.[22] (We discussed such sets in Chapter 3 with reference to Cantor's theorem.) For such sets, it seems quite reasonable to assume that choice holds. Indeed, Specker's proof shows it failing

[21] Adolf Lindenbaum and Alfred Tarski, "Communication sur les Recherches de la Théorie des Ensembles [1926]," in Alfred Tarski, *Collected Papers*, vol. I, eds. Steven R. Givant and Ralph N. McKenzie (Boston, MA: Birkhäuser, 1986), pp. 171–204. They sketch their result on pp. 187–8. Waclaw Sierpinski proved it in "L'hypothèse généralisée du continu et l'axiom du choix," *Fundamenta Mathematicae* 34 (1947), pp. 1–5.

[22] Rosser, "Review of Specker," p. 127.

only for non-Cantorian sets such as the universal set.[23] Quine says that we might think of the situation as follows:

> So what we have is just another case, and a beauty, of unconventional behavior on the part of non-Cantorian classes ... Classical results can still be got in NF by inserting, where needed, a premiss that a class concerned is Cantorian. One could look upon NF as merely more general, in this respect, than set theories where everything is Cantorian.[24]

Furthermore, the possibility that these non-Cantorian sets might yield important results for the Cantorian sets should not be overlooked:

> The demonstrability of "$\Lambda \notin N$" [a version of the axiom of infinity] suggests that non-Cantorian classes, with all their unconventional behavior, are not just a harmless annex to an otherwise well-behaved universe, but a valuable intermediary in proving desirable theorems about the well-behaved objects themselves. For it is the failure of the axiom of choice for non-Cantorian classes that has enabled Specker to prove "$\Lambda \notin N$", which itself says nothing of non-Cantorian classes.[25]

Quine's remark here is particularly interesting because it seems rather close to the kind of reasoning that is often provided by set theorists working in the ZF tradition trying to justify large cardinal hypotheses. While such cardinal numbers might initially appear too large to have anything to do with ordinary mathematics – never mind natural science – it turns out that they often do have implications for the more ordinary, undisputed, lower-level mathematics. Quine is here suggesting an analogous situation with regard to the big sets of NF such as the set of cardinal numbers, the set of ordinal numbers, and the universal set itself.[26]

As I have already remarked, mathematicians are generally careful about their use of choice, not necessarily doubting its truth but still keeping track of where it is necessary to a result.[27] Rosser is no exception. Even

[23] We should also note that NF is not the only set theory for which AC fails. The axiom of determinacy has often received serious consideration as an additional axiom to add to ZF-style set theories, but as with NF, such theories also disprove AC. For more on this topic, see Akihiro Kanamori, *The Higher Infinite*, 2nd edn (New York, NY: Springer, 2003), pp. 377–8.

[24] Quine, *Set Theory and Its Logic*, p. 296.

[25] Ibid., p. 299.

[26] Although he does not cite this comment from Quine himself, Thomas Forster also makes this point in his *Stanford Encyclopedia of Philosophy* entry "Quine's New Foundations": https://plato.stanford.edu/entries/quine-nf/. For some indication of the way in which large cardinal hypotheses have consequences for lower level mathematics, see Hrbacek and Jech, *Introduction to Set Theory*, pp. 282–3.

[27] In a way, Quine's work in *Set Theory and Its Logic* can be viewed as a generalization of this approach to set theory.

before Specker's proof, when it might have seemed completely reasonable to assume the full axiom of choice, that is, choice for the universe, Rosser had already limited himself in his 1953 *Logic for Mathematicians*, which takes NF as its basis, to the denumerable axiom of choice: the existence of a choice function for countable families of nonempty sets.[28] Indeed, denumerable choice suffices for the development of most mathematics, including that of the real numbers.[29] As Rosser describes the situation in a post-Specker appendix to *Logic for Mathematicians*, there are an infinite number of choice axioms, each of which says that we can make λ many choices, where λ is any cardinal number. It is only where λ is the cardinality of the universe that Specker's proof holds.[30] Hence this result does not threaten Rosser's reduction of mathematics to NF.

The failure of choice for NF generally may still seem to be an oddity, perhaps an uncomfortable one, since appeal to it has become so ingrained in contemporary mathematical practice. While Zermelo's well-ordering theorem claimed to show that all sets can be well ordered, the theorem cannot hold in NF, since this is equivalent to the axiom of choice. But we should not forget that Zermelo's set theory rules out certain sets from the start, most noticeably – in comparison to NF – the universal set. And the universe is such a set that is non-well-orderable.[31] If our starting point does not allow for such sets, then there are no problems for Zermelo's well-ordering principle. But the very question that I have been trying to raise throughout this book is about what grounds there are for ruling out sets such as the universal set from the start. It seems to have a perfectly clear specifying condition, and it also seems perfectly reasonable that in the course of set-theoretic research (because our interest is in sets as such), we might want to know what the universe and other big sets are like. Indeed, as we have seen, this was one of Quine's driving motivations in proposing NF in the first place. With the proof of the axiom of infinity, we saw an example of how these big sets might yield insights into the more conventional sets that are present in the more common axiomatizations of set theory. Why not just do as Quine urged, then, and think of NF as merely a more general theory, one that includes sets that ZF leaves out?

[28] J. Barkley Rosser, *Logic for Mathematicians*, 2nd edn (Mineola, NY: Dover Publications, 1978), p. 512.

[29] For some discussion and examples, see Potter, *Set Theory and Its Philosophy*, pp. 161–5.

[30] Rosser, *Logic for Mathematicians*, pp. 540–1.

[31] However, the universe can be well ordered in NFU as long as there are enough atoms (urelements) in the universe. For details, see Randall Holmes "The Set-Theoretical Program of Quine Succeeded, but Nobody Noticed," *Modern Logic* 4 (1994), pp. 1–47.

It should be noted that NF is not the only set theory to disprove the axiom of choice. This occurs also in ZF-style set theories to which the axiom of determinacy has been added. To gain some understanding of the axiom, start by thinking of a real number as being represented by an infinite sequence of 0s and 1s.[32] A pair of players in an infinite game then take turns playing a 0 or a 1, which will generate a real number. Now, if *A* is a set of real numbers, there is a game *G*(*A*) associated with this set. The first player will win the game if the real number generated by the two players is in *A*, and the second player will win if it is not. This game and the set *A* are said to be *determined* if and only if one of the players has a winning strategy for the game. The axiom of determinacy (AD) states that every set of real numbers is determined.[33] This axiom might seem to lack much in the way of intuitive motivation as a natural candidate for being added to the axioms of ZF. But by a theorem of Jan Mycielski and Stanislaw Swierczkowski, it does imply some very desirable properties for sets of real numbers, namely, that every set of real numbers is Lebesgue measurable, has the Baire property, and has the perfect set property.[34] However, AD also contradicts the axiom of choice, since choice implies the existence of an undetermined set of real numbers. This was, however, much the point; AD was supposed to eliminate the undesirable nonmeasurable sets that result from full choice.[35] So here we have an additional axiom candidate for ZF that, like the system of NF as a whole, denies the axiom of choice. In recent times, largely owing to its denial of choice, AD has received less attention as a plausible additional axiom for ZF. However, in an early monograph on AD, Eugene Kleinberg argues in much the same way that I argued with regard to the failure of choice in NF:

> Contradicting the axiom of choice is not, in and of itself, grounds for rejecting the axiom of determinateness. Set theorists long ago segregated the axiom of choice from the standard axioms of set theory due to its

[32] In my exposition here, I follow that found in Penelope Maddy, *Naturalism in Mathematics* (Oxford: Clarendon Press, 1997), p. 79. Hrbacek and Jech's account in their *Introduction to Set Theory*, pp. 280–1, is more detailed and slightly more technical but is also quite accessible.

[33] There are variations in how the axiom is stated. Kanamori also states the axiom this form in his *The Higher Infinite*, p. 377. Chapter 6 of his book provides a thorough technical treatment of determinacy as well as its historical setting. For further details, see also Yiannis Nicholas Moschovakis, *Descriptive Set Theory* (New York, NY: North-Holland, 1980), chapter 6.

[34] This is stated in Kanamori, *Higher Infinite*, p. 377, along with some useful historical context. The original paper is Jan Mycielski and Stanislaw Swierczkowski, "On the Lebesgue Measureability and the Axiom of Determinateness," *Fundamenta Mathematicae* 54 (1964), pp. 67–71. It should be noted that in earlier discussions of AD, it tended to be called the axiom of determinateness.

[35] Kanamori, *The Higher Infinite*, pp. 377–8.

> somewhat tenuous nature. Mathematicians in general had been aware of problems with the axiom of choice, and following such things as the Banach-Tarski paradox, analysts avoided it whenever possible.[36]

Most research on determinacy has now shifted to the axiom of projective determinacy, which still yields the desirable properties for sets of real numbers but does not deny the axiom of choice. In a further connection with NF, AD does still allow for weaker forms of choice. Rather than denying AD, Jech, in his standard work on set theory, adopts this strategy of assuming a weakened form of choice.[37] Because AD has large cardinal consequences, one might start to wonder whether there is just no reason to expect choice to hold as we get closer and closer to obtaining the entire universe of set theory.[38]

As a final point on the axiom of choice, recall Boolos's comment about an ideal theory: that it should decide the continuum hypothesis, which is independent of the axioms of ZF (and ZFC). Choice is also independent of ZF and has had a controversial history with regard to its truth.[39] Most logicians and mathematicians now accept it but largely, as we have seen, because of its pragmatic value for mathematics overall; it yields many desirable results and no (or perhaps not many) undesirable ones. But NF, too, can have choice – just not in the unrestricted form in which ZFC has it. Again, this is because ZFC is already restricted in its banning of such sets as the set of cardinal numbers, the set of ordinal numbers, and the universal set itself. Restricted versions of choice, such as the denumerable axiom of choice or choice restricted to Cantorian sets, will work perfectly well in NF and provide for much of ordinary mathematics, just as ZFC does. Yet NF is perhaps more ideal (in Boolos's sense) than ZF with regard to choice. Choice is independent of ZF and so must be added as an additional axiom if ZF is to serve as a framework for mathematics. In a sense, then, ZF gives us no reason to think that choice is true. Choice yields many desirable consequences, so it makes sense that we would want it to be true. But wanting something and its actually being so are two different things. NF tells us that choice in its most general form is false and so actually decides a set-theoretic statement that ZF does not. This does

[36] Eugene M. Kleinberg, *Infinitary Combinatorics and the Axiom of Determinateness*. Lecture Notes in Mathematics 612 (New York, NY: Springer-Verlag), p. 3.

[37] Thomas Jech, *Set Theory*, 3rd edn (New York, NY: Springer-Verlag, 2003), p. 628.

[38] Potter, *Set Theory and Its Philosophy*, pp. 278–9, includes a discussion of the point at which, in our assumption of larger and larger large cardinal axioms, we have to give up choice.

[39] Again, see Moore's *Zermelo's Axiom of Choice* for a complete history of the axiom, including the controversies that it raised.

not inhibit the development of mathematics in NF, since choice holds for Cantorian sets. Indeed, we might think of this failure as telling us something further about *all* the sets there are. In ZF-style set theory, we build in the assumption that all sets are Cantorian. With NF, we explore the set-theoretic universe as a whole without prejudice toward particular kinds of sets. In the failure of choice, we then gain some insight into the general properties of non-Cantorian sets and what distinguishes them from Cantorian sets.[40] Furthermore, as a consequence of disproving choice, we also gain information about the generalized continuum hypothesis (GHC). A result of Tarski and Adolf Lindenbaum shows that GHC proves choice, so GHC is also false in NF.[41] So in Boolos's sense, it seems that NF may be a more ideal theory than ZF (or ZFC) in its ability to decide questions such as these.

7.2 Arithmetic in New Foundations

Let me return now to more mundane matters of arithmetic. That the failure of choice in NF does not inhibit the development of ordinary mathematics in this theory is reassuring, but this might not be the only worry to have. As is well known, developing mathematics set-theoretically generally proceeds by first showing that set theory captures arithmetic. The other number systems of integers, rational numbers, and real numbers are then built up from this basis. So we also want to be certain that NF can provide for arithmetic, and here, further difficulties arise. Because NF's comprehension schema is restricted by stratification, certain inductive definitions that are used in arithmetic will be problematic. In this sense, then, despite allowing for big sets, NF is more restrictive than ZF.[42]

[40] Forster, in his "The Status of the Axiom of Choice in Set Theory with a Universal Set," *Journal of Symbolic Logic* 50:3 (1985), pp. 701–7, tries to sketch out some general characteristics of where and why AC fails for set theories with a universal set. There are, of course, theories related to ZF for which AC fails, e.g., ZF + Determinacy. It seems to me that this may be one particularly good reason for taking research on NF seriously: Perhaps we will find trends across set theories that will help us to see connections rather than differences – in a sense, to come to a better understanding of a unified body of knowledge called set theory rather than ZF-style set theory (which is usually taken to be set theory simplicter) and its deviant relatives such as NF. This seems to be perfectly in line with a Quinean approach to set theory as a comparative endeavor investigating whatever sets there might be rather than just our favorite ones. In *Set Theory and Its Logic*, Quine himself describes the failure of choice not as a mark against NF but rather as "just another case, and a beauty, of unconventional behavior on the part of non-Cantorian classes … One could look upon NF as merely more general, in this respect, than set theories where everything in Cantorian" (p. 296).

[41] Lindenbaum and Tarski, "Communication sur les Recherches de la Théorie des Ensembles, and Sierpinski, "L'hypothèse généralisée du continu et l'axiom du choix."

[42] NF is in much the same situation as type theory on this point.

For NF, sets cannot be specified when their specifying conditions are unstratified, and this comes into play with induction, since we will then have induction only for stratified conditions. So, for example, we might expect that we should be able to prove for each natural number n that it counts the set of its predecessors.[43] That is, we might expect to prove by induction $(\forall n)(n \in \mathrm{N} \supset \{m: m \in \mathrm{N} \,.\, 0 < m \leq n\} \in n$, but we see that this condition is not stratified, since n occurs on both sides of the membership relation. Therefore we will not be able to state the existence of such a set in NF.

To resolve this sort of issue, Rosser put forward the axiom of counting, which states that for any natural number n, n counts the number of its predecessors, $(\forall n \in \mathrm{N})\{m: 0 < m \leq n\} \in n$. Equivalent to the axiom is the type-raising operation T for ordinals (analogous to T for cardinals) when it is assumed that T is the identity map when applied to natural numbers, that is, for all natural numbers n, $T\{n\} = n$, or the order-type of the relational unit subsets of n is equal to n. This further yields that all finite sets are Cantorian, that all finite sets are strongly Cantorian (that is, the singleton function is a set), and that there is an infinite strongly Cantorian set. Indeed, these claims are equivalent under the axiom of counting.[44] The axiom of counting also helps to preserve a variety of more ordinary mathematical results, such as the theorems of number theory that concern the number of integers having a particular property. It also simplifies the natural numbers generally in assuring us that if n is a natural number, then for any set $\alpha \in n$, α has n members.[45]

But let us not forget that counting is an additional axiom that is not deducible from NF's two original axioms: extensionality and comprehension.[46] We might try to justify it on the basis of its rather intuitive-sounding nature, though we should be wary of such justifications in past

[43] The natural numbers in NF are of the Frege-Russell variety, that is, 0 is the set of all sets with no members, 1 is the set of all singletons, 2 is the set of all pairs, and so on.

[44] For details, see Forster, *Set Theory with a Universal Set: Exploring an Untyped Universe* (Oxford: Clarendon Press, 1992), pp. 30–1; and Rosser, *Logic for Mathematicians*, pp. 485–7.

[45] For these and some other examples of counting in ordinary mathematics, see Rosser, *Logic for Mathematicians*, pp. 485–9.

[46] Quine's system Mathematical Logic – NF with proper classes added – does not face the problem with induction but requires an additional axiom to guarantee that the natural numbers form a set. Quine thought such an addition inelegant; see his "The Inception of 'New Foundations'," in *Selected Logic Papers*, enlarged edn (Cambridge, MA: Harvard University Press, 1995), p. 289. No doubt he would think the same about adding the axiom of counting to NF. He states this for supplementing NF in other ways to secure induction; see *Set Theory and Its Logic*, p. 297. This sort of dissatisfaction seems just to bring us back to Zermelo's observation that no solution to the paradoxes has been found that is "entirely satisfactory." We will always be giving up something of our most natural understanding of sets.

work, for example, with regard to choice. Indeed, much like choice, counting is a surprisingly powerful axiom. In fact, NF plus counting (NFC) proves the consistency of NF, so on pain of inconsistency (by way of Gödel's second incompleteness theorem), it had better not be the case that counting follows from the two initial axioms. Still, we might say that counting fares somewhat better than choice in not yielding any especially counterintuitive results – aside from those that are already present in NF without counting (and these results tend to come from the presence of non-Cantorian or big sets in NF and so are perhaps counterintuitive only because other set theories do not allow us to investigate such sets). Unless we demand some further self-evident grounding for our mathematics, we seem to be no worse off in our choice of axioms than we are with ZF (or ZFC). Indeed, in its economy of axioms, NF recommends itself considerably over ZF (and ZFC). This is enough for the point I have been arguing: that, so far at least, the view that ZF somehow gives us the only viable account of sethood relies more on prejudice than on fact.

Despite being surmountable, the difficulty over induction that emerges out of NF's restriction to stratified conditions suggests that perhaps NF should not be preferred as a foundation for ordinary mathematics. This is no objection to my view of NF. My aim here has only been to show that, like other set theories, NF can capture ordinary mathematics and that, also like other set theories, its axioms will need supplementing to do so.[47] Even if we should still think that NF develops arithmetic in a way that is much more unnatural than the way in which it is developed in, say, ZF, this is just a tradeoff between the two set theories. What NF gives us in return that ZF does not is a way into the mathematics of the absolute infinite. This is a point to which I will return shortly.

7.3 Exploring the Set-Theoretic Universe

In this concluding section, I want to try to provide some summary of what exactly a Quinean philosophy of set theory looks like.[48] Given what we have seen of its exploratory and experimental nature, its purpose cannot be

[47] Again, other set theories require an axiom of infinity, whereas NF requires the axiom of counting.

[48] Joseph Ullian, in his overview of Quine's work in logic and its relevance to his philosophy, has influenced my understanding of Quine's logical work greatly, not only in what follows immediately but throughout this book. See his "Quine and the Field of Mathematical Logic," in *The Philosophy of W. V. Quine*, eds. Edwin Hahn and Paul Schilpp, expanded edn (Chicago, IL: Open Court Publishing, 1998), pp. 569–89.

to bring forth something like the essence of sethood as a sort of a priori science. Quine is a pluralist about set theory, as I take it he is in some sense a pluralist about all our scientific theories. This is not to say that almost anything might count as set theory for him. Indeed, the point of Section 7.2 was to show, according to some general and rather uncontroversial criteria, that Quine's NF could be reasonably included within the discipline of set theory. So, given his approach to set theory, I want to return to the topic of what a Quinean philosophy of set theory is, if not to bring forth the essence of sethood, for example, in terms of an iterative hierarchy.

I argued in Chapter 4 that the culmination of Quine's work in set theory came in his 1963 *Set Theory and Its Logic*, a work that, I claim, stands alongside *Word and Object* as one of his greatest philosophical achievements.[49] *Set Theory and Its Logic* is dedicated to Russell for initiating Quine's interest in the subject, and it has as its epigraph a quote from Gilbert and Sullivan's *The Pirates of Penzance*: "How quaint the ways of paradox," stressing again the central importance Quine sees in the paradoxes as giving rise to the development of set theory. His major theme is a comparative study of the main approaches to set theory. As we saw in the first two parts of the book, Quine develops a set theory that he intends as largely neutral between the usual systems of Russell, Zermelo, and himself. He describes his policy as one of minimizing assumptions about set existence, making them only where the development of the theory demands them.[50] To this end, he employs the apparatus of virtual classes, simulating talk of classes by way of contextual definitions rather than assuming their existence from the start. This gives him a useful contrast later, when he introduces actual classes, better demonstrating "what power real classes confer that the counterfeits do not."[51] He adopts axioms that are strong enough only to imply the existence of finite sets. Where more substantial claims enter in, such as the existence of infinite sets or the axiom of choice, he adopts these only as hypothetical claims at the outset of proving the theorem that requires these added assumptions.[52] It is only after he has introduced his neutral and largely minimal set theory that he thinks we are ready, in the third part of the book, to consider the relative advantages and disadvantages of

[49] Quine, *Set Theory and Its Logic*; W. V. Quine, *Word and Object* (Cambridge, MA: MIT Press, 1960).
[50] Quine, *Set Theory and Its Logic*, p. vii.
[51] Ibid., p. ix.
[52] For example, see Quine's discussion of this approach in *Set Theory and Its Logic*, p. xi.

the various more familiar and substantial set theories. His aim is to prepare us to consider the relative merits of the usual set theories from a perspective that is prejudiced toward none of them.[53]

Quine also returns to the theme that "intuition is bankrupt" when it comes to developing a viable system of sets, for intuition brought set theory in its earliest days to contradiction. This is not to say that intuitions never come into play or are never helpful in set theory, but we need to be cautious, as Quine explains (I put forth a similar argument in Chapter 6, but here we have Quine himself making it):

> The notion of class is so fundamental to thought that we cannot hope to define it in more fundamental terms. We can say that a class is any aggregate, any collection, any combination of objects of any sort; if this helps, well and good. But even this will be less help than hindrance unless we keep clearly in mind that the aggregating or collecting or combining here is to connote no actual displacement of the objects, and further that the aggregation or collection or combination of say seven pairs of shoes is not to be identified with the aggregation or collection or combination of those fourteen shoes, nor with that of the twenty-eight soles and uppers. In short, a class may be thought of as an aggregate or collection or combination of objects just so long as "aggregate" or "collection" or "combination" is understood strictly in the sense of "class."[54]

This is not to say that we have no way of knowing what a class is. Quine does think that there is some use for coming to understand classes by way of an analogy to the ordinary collections of physical objects we come in contact with every day. But to understand collections *as classes*, he turns to predicates. If we begin with a sentence about a thing, we can then think of removing any reference to that thing from the sentence, leaving us with only the predicate, or an open sentence, true of some things and false of others. The class, then, as we now know, is exactly those things for which the predicate is true – the extension of the predicate. Quine also notes that we want classes to be identical when they have the same members, so we also adopt extensionality as our principle of class identity. But he continues to be careful in appealing to intuitions; his gesturing at what a class is as distinct from ordinary collections is not meant to be definitive:

> I was describing the function of the notion of class, not defining class. The description is incomplete in that a class is not meant to require, for

[53] As Quine remarks, on p. x of *Set Theory and Its Logic*, "Because the axiomatic systems of set theory in the literature are largely incompatible with one another and no one of them clearly deserves to be singled out as standard, it seems prudent to teach a panorama of alternatives."

[54] Quine, *Set Theory and Its Logic*, p. 1.

> its existence, that there be an open sentence to determine it. Of course, if we can specify the class at all, we can write an open sentence that determines it … But the catch is that there is in the notion of class no presumption that each class is specifiable. In fact there is an implicit presumption to the contrary, if we accept the classical body of theory that comes down from Cantor.[55]

Maintaining Cantor's theory of the infinite is not the only reason Quine cites for not adhering too closely to the idea of classes as the extensions of predicates. Because this "natural attitude," as he describes it, led to contradiction, we must instead make deliberate and careful choices about our class existence axioms, since "intuition is not in general to be trusted here." But because we have a variety of interesting alternative axiomatizations available, Quine thinks that it hasty to focus on just one of them "to the point of retraining our intuition to it."[56] In the contemporary philosophy of set theory, this is exactly what has happened. Being a viable set theory has become nearly inseparable from the intuitions guiding the iterative conception of set.[57]

In Quine's view, then, seeming artificiality is not a reason to reject a particular set theory. Yet he does not rule out the possibility that some particular set theory will eventually be adopted as best, for he thinks that his own approach – a careful weighing of the benefits and drawbacks of the various options – "can encourage research that may some day issue in a set theory that is clearly best."[58] As we might expect, his attitude here is much like the attitude one could have with regard to the progress of science generally. Various theories are developed until one emerges as best, however temporary this privileged position might happen to be.

This careful comparative development of set theory with regard to its existence assumptions and axiomatic strength is, for Quine, the philosophy of set theory. This should come as no surprise to anyone who is familiar with his approach to the philosophy of the sciences generally. Neither in the natural sciences nor in mathematics do we look for a first philosophy. Our starting point is within science itself, so in the case of set theory, its philosophy begins within set theory rather than in some extramathematical metaphysical conception to which it must be fitted.

[55] Ibid., p. 2.

[56] Ibid., p. 5.

[57] Something like this seems to be the source of the dispute that arose between Quine and Donald Martin over the latter's review of *Set Theory and Its Logic*. See Martin's "Review of *Set Theory and Its Logic*," *Journal of Philosophy* 67:4 (1970), pp. 111–14, and Quine's "Reply to Martin," *Journal of Philosophy* 67:8 (1970), pp. 247–8.

[58] *Set Theory and Its Logic*, p. x.

Having cleared his readers' minds of set-theoretic prejudice in the first two parts of *Set Theory and Its Logic*, Quine considers in the third part the familiar set theories of Russell, Zermelo, and himself. Rather than drawing distinctions and building boundaries between them, Quine takes the very radical approach of bringing these various and often incompatible set theories into dialog with each other. So let us turn to the details of Quine's picture and how it might further the growth and knowledge of set theory.

Quine begins his study with Russell's theory of types, progressing next to Zermelo's theory and finally to his own NF. This account is not historical; nor is it intended to be. Rather, we can view Quine's account as a sort of logical progression through the various set theories that he considers, showing how each one can be seen, in a sense, as emerging from the other. Boolos and Russell, among others,[59] argue against Quine's NF by remarking on its seemingly artificial development as only a response to the paradoxes, asserting that there is no motivating thought behind it other than this. Quine largely accepts this but sees it as the situation generally in set theory after the paradoxes. To him, it is not a criticism of NF any more than it is a criticism of the other set theories that he considers. The paradoxes force such artificiality upon all of us. As we have heard over and over from Quine, our most intuitive thought about sets, that is, as extensions of predicates, led us into contradiction, and we have seen this play out in the early history of set theory. Talk of sets existing in some sort of hierarchy did not enter in as an intuitive first thought about sets but rather emerged from considerations of paradoxes, first in Russell's work and sometime later in Zermelo's (much later than his first axiomatization). Artificial development, then, is unavoidable.

Perhaps critics of NF would grant Quine's point that all set theory is in part artificial, but for them, NF is just a little too artificial in comparison to the theory of types or Zermelo's set theory. Quine's comparative account of set theory can be read as a response to just this sort of criticism. Aside from comparing the relative merits of the various set theories and how their respective disadvantages may lead one to a different axiomatization, Quine – perhaps even more importantly – brings out what unifies them. To this end, he begins with types and shows how its hierarchical structure leads to Zermelo's theory and then finally to his own NF. What Quine does is take this talk of hierarchy and abstract away

[59] I should include here Martin's "Review of *Set Theory and Its Logic*." See also Quine's "Reply to Martin."

from this metaphor as much as possible until he is left with its barest logical structure, present across the boundaries of the various competing and incompatible axiomatizations of set theory. This comparison accounts for certain intuitions about set theory while also moving against our becoming too attached to certain intuitions that we might have about set theory. The final section of *Set Theory and Its Logic*, then, is a crucial move in an argument against those who would focus on just one formulation of set theory "to the point of retraining our intuition to it."[60]

Let us see now how this argument goes. We discussed in Chapter 2 how Russell saw his theory of types as emerging from Henri Poincaré's diagnosis that the contradictions all have their origins in quantifying over illegitimate totalities, which Russell took to be summed up in his vicious circle principle. Recall that in its simplified version, the theory of types divides the universe into levels: individuals at the lowest level, then classes of individuals, then classes of classes of individuals, and so on. For something to be a member of a class, it must be of level n while the class of which it is a member is of level $n + 1$. This dividing of the universe into levels, motivated by Russell's vicious circle principle, apparently restored consistency to the theory of classes. But Quine observes that this move is drastic in that its type restrictions tamper with the original logic. Furthermore (as we saw in Chapters 3 and 4), type theory has the undesirable feature of reduplicating objects in other levels of the hierarchy, among them the empty class, a series of quasi-universal classes, and the various objects of mathematics such as the different number systems.

We might take these drawbacks alone as moving us beyond types to some other approach to set theory, but Quine observes that there is also something valuable in the constructive metaphor of type theory: "it is a part of set theory that carries extra conviction, because of the construction metaphor."[61] Quine then moves to Zermelo set theory with its unrestricted (with regard to type) variables not merely as a reaction to the undesirable features of types but rather as a natural generalization of types' hierarchical structure. Whereas other philosophers of set theory have ruled out types as set theory, often by fiat, Quine engages with both theories to demonstrate how we might see the interconnections between them.[62] Let us consider

[60] *Set Theory and Its Logic*, p. 5.

[61] Ibid., p. 264.

[62] The standard move is to declare types to be part of logic, that is, higher-order logic, and so not of set theory. Quine remarks on this in *Set Theory and Its Logic*, pp. 257–58, and in many other places in his writings. Boolos makes this sort of declaration in his 1975 "On Second-Order Logic," in *Logic, Logic, and Logic*, p. 42, fn. 9, in discussing Quine's presentation of a class theory in the third edition of his *Methods of Logic* (Chicago, IL: Holt Rinehart, and Winston, 1972).

Quine's approach by again looking at how Zermelo's set theory emerges in a natural way from Russell's theory of types.

To show how Zermelo's theory emerges from Russell's, Quine first observes that typical ambiguity in types does not change a many-sorted theory into one with general variables. The point of typical ambiguity is that although we often do not explicitly specify the types involved for each formula (for ease of writing and reading), we must still be certain that they can be made explicit in the appropriate way if we are called upon to do so. So, for example, the typically ambiguous formula "$(\exists y)(\forall x)(x \in y)$" can be shown to be legitimate by restoring type indexes according to the scheme "$(\exists y^{n+1})(\forall x^{n})(x^{n} \in y^{n+1})$." However, Quine observes that we may also have reasons to consider the translation of the many-sorted types into a theory of general variables. For instance, the type indices are themselves cumbersome, but typical ambiguity can make things worse in some cases. Consider the two formulas "$x \in y$" and "$y \in x$." Both are meaningful under types, yet "$x \in y \,.\, y \in x$" is not. In this way, we see that the theory of types blocks the set-theoretic contradictions with the drastic move of revising general logic. Quine, favoring the maxim of minimum mutilation, thinks that this is to be avoided where possible. Furthermore, given his comparative interest in set theory (and scientific theories generally), a single underlying logic makes such studies far more feasible.

Quine argues that the many-sorted logic is not essential to type theory. Instead, we can allow the variables to be completely general and add a predicate T_n to impose the type restrictions. Formulas can then be restricted in the familiar way as follows: "$(\forall x)(T_n x \supset Fx)$" or "$(\exists x)(T_n x \,.\, Fx)$." This move, Quine says, makes sense of Russell's grammatical restrictions.[63] Whereas Russell declared all formulas of the form "$x^m \in y^n$" with $n \neq m + 1$ meaningless, we can now take such formulas to be simply false. Therefore the grammatical restrictions generated by types appear unnecessary in the first place. The two axiom schemas of types go over into the new theory as follows: Comprehension and extensionality become "$(\exists y)(T_{n+1}y \,.\, (\forall x)(T_n x \supset x \in y \equiv Fx)$" and "$(T_{n+1}x \,.\, T_{n+1}y \,.\, (\forall w)(T_n w \supset (w \in x \equiv w \in y))) \supset x = y$," respectively. (Officially, identity is defined in terms of membership and first-order logic and so is eliminable.)

[63] This was one of Quine's earliest motivations for his attempts to rework Russell's logic. See Quine's remarks on his 1936 "Set-Theoretic Foundations for Logic," in his "The Inception of New Foundations" in *Selected Logic Papers*, pp. 286–7.

Having shown the eliminability of typed variables in favor of typing predicates, Quine goes on to show how smoothly types can lead us into Zermelo's set theory by converting the restricted variables of types into cumulative types with general variables.[64] We begin by equating the empty sets of all the various types so as to get a single empty set. Quine then introduces the trick of identifying individuals, the memberless objects of type 0,[65] with their unit classes and treating membership between individuals as identity. The objects of type 0, the T_0x's – the individuals – can now be defined by the formula "$(\forall y)(y \in x \equiv y = x)$" and the objects of higher type, the $T_{n+1}x$'s – the classes – by the formula "$(\forall y)(y \in x \supset T_n y)$." Any typed variable can be rewritten according to these definitions so that we use only completely general variables, and the types become cumulative, that is, $T_m x \supset T_n x$ for all $m < n$.

Here, we see just the sort of philosophy of set theory in which Quine engages. Rather than looking at differences and attempting to privilege one set theory over another, he tries to see their similarities. For Quine, coming to understand the realm of sets means investigating sets from the various perspectives that different set theories allow for. His endeavor in set theory, as in much of science, is cooperative rather than exclusionary and aims to broaden our knowledge through pluralism about set theory. This moving from noncumulative to cumulative types perhaps allows us to better understand the apparent hierarchical structure of sets. It was always present in both types and Zermelo set theory, and Quine has now shown us explicitly how the idea connects the two approaches to set theory. Quine himself found this connection immensely striking. He later observed of Zermelo's theory that "in its multiplicity of axioms it seemed inelegant, artificial, and ad hoc. I had not yet appreciated how naturally his system emerges from the theory of types when we render

[64] The details can be found in *Set Theory and Its Logic*, pp. 272–4. Here, I hope to give only the basic idea behind the conversion to cumulative types and Zermelo set theory. This was an idea that Quine was already onto much earlier, which he laid out in his "Unification of Universes in Set Theory," *Journal of Symbolic Logic* 21:3 (1956), pp. 267–79. It is worth noting that Quine here sees types and Zermelo's theory as sharing more with each other than with NF, and although he states a preference for his own theory, he also observes that "in the present paper I am pursuing other lines in order to dramatize the process of unifying the universes of theories containing multiple styles of variables" (p. 275). Here, we see the distinct approach that Quine takes to set theory. While so much of the philosophy of set theory concerns singling out a privileged theory, Quine instead looks for where they might be of a piece with one another.

[65] Individuals are necessary to types in accounting for the infinite. The axiom of infinity for types says that there are an infinite number of individuals.

the types cumulative and describe them by means of general variables."[66] So while many have remarked on the artificiality of Quine's systems, Quine initially thought the same of Zermelo's. The sort of elegance, simplicity, and unity that Quine (like many other philosophers of science) has so often put forth as desirable for scientific theories generally[67] comes now also to Zermelo's theory by way of its emergence from the perhaps more intuitive theory of types. There are, of course, a variety of ways in which we might consider types to be intuitive. Here, I have in mind (as I suspect Quine does) that types, at least initially, seemed to do the least damage to the original conception of sets as extensional and specified by a comprehension principle of a sort. In linking these theories, this sort of intuitiveness of types passes on to Zermelo. Indeed, Quine concludes that had he appreciated this link earlier, "I might not have pressed on to 'New Foundations.'"[68]

Bringing out such connections now also gives us another way to see NF as sharing an important aspect with types and Zermelo set theory. The move from types to Zermelo's theory perhaps gives us further reason to consider the idea of a hierarchy as somehow essential to what a set is. But as we saw in Chapter 6, it is notoriously difficult to make philosophical sense of this idea. On the one hand, we have talk of sets coming into being at particular levels of the hierarchy; on the other hand, we have talk of sets existing eternally from the past and to the future. With NF, Quine perhaps provides a way of making sense of this idea of hierarchy while freeing it from a particular metaphysical account of sets in the following way. We can think of NF as abstracting away completely from the idea of a hierarchy until we are left with a much weaker purely syntactic account of set existence. As Burton Dreben has put it – and Joseph Ullian, following him – Quine's approach is one of syntactic exploration.[69] The hierarchy gives us only the numbering scheme that is used as NF's syntactic test for set existence, an idea that Quine came upon by taking typical ambiguity very seriously, which appears to have no need for an actual hierarchy of sets.[70] We need not make the further

[66] Quine, "The Inception of New Foundations" (1987), in *Selected Logic Papers*, enlarged edn (Cambridge, MA: Harvard University Press 1995), p. 287.

[67] See, for example, Quine's "Posits and Reality," in *The Ways of Paradox and Other Essays*, pp. 247–8.

[68] Quine, "The Inception of New Foundations," p. 287.

[69] Ullian, "Quine and the Field of Mathematical Logic," pp. 584 and 588, n. 56.

[70] Quine, *Set Theory and Its Logic*, pp. 287–9. See also his remarks in "The Inception of New Foundations," p. 244, where he discusses how he arrived at NF by considering Russell's types in conjunction with Zermelo's theory.

leap to, say, the iterative hierarchy to get an account of the essence of sethood. This would be an extra, more robust philosophical assumption. Quine's syntactic account instead keeps our feet on the ground.[71] In doing the philosophy of set theory, Quine tries to keep to his naturalistic strictures, and we must always remember that for him, "philosophy of science is philosophy enough."[72]

[71] Quine makes this remark in his *Methods of Logic*, 4th edn, p. 212, when considering the substitutional versus set-theoretic account of logical truth. The point seems apt here as well: We try to get by with fewer assumptions where possible.

[72] W. V. Quine, "Mr. Strawson on Logical Theory," in *The Ways of Paradox and Other Essays*, p. 151.

Bibliography

Aczel, Peter. *Non-Well-Founded Sets* (Stanford, CA: CSLI Publications, 1988).

Banach, Stefan and Alfred Tarski. "Sur la Décomposition des ensembles de points en parties respectivement congruents" (1924). In Tarski, *Collected Papers*, pp. 121–54.

Barrett, Robert B. and Roger Gibson, eds. *Perspectives on Quine* (Cambridge: Basil Blackwell, 1990).

Barwise, Jon, ed. *Handbook of Mathematical Logic* (New York, NY: North-Holland, 1977).

Barwise, Jon and John Etchemendy. *The Liar: An Essay on Truth and Circularity* (New York, NY: Oxford University Press, 1987).

Benacerraf, Paul and Hilary Putnam, eds. *Philosophy of Mathematics: Selected Readings*. 2nd edn (New York, NY: Cambridge University Press, 1983).

Boolos, George. *Logic, Logic, and Logic*. Ed. Richard Jeffrey, with Introductions and Afterward by John P. Burgess (Cambridge, MA: Harvard University Press, 1998).

"On Second-Order Logic" (1975). In *Logic, Logic, and Logic*, pp. 37–53.

"The Iterative Conception of Set" (1971). In *Logic, Logic, and Logic*, pp. 13–29.

Burali-Forti, Cesare. "A Question on Transfinite Numbers" (1897). In van Heijenoort, *From Frege to Gödel*, pp. 104–11.

"On Well-Ordered Classes" (1897). In van Heijenoort, *From Frege to Gödel*, pp. 111–12.

Cantor, Georg. *Contributions to the Founding of the Theory of Transfinite Numbers (1895, 1897)*, trans. Philip E. B. Jourdain (New York, NY: Dover, 1955).

Gesammelte Abhandlungen: Mathematische und Philosophischen Inhalts. Ed. Ernst Zermelo (Hildesheim: Georg Olms Verlagsbuchhandlung, 1962).

"Mitteilungen zur Lehre vom Transfiniten" (1887, 1888). In *Gesammelte*, pp. 378–439.

"On a Property of the Totality of All Real Algebraic Numbers" (1874). In Ewald, *From Kant to Hilbert*, pp. 839–43.

"On an Elementary Question in the Theory of Manifolds" (1891). In Ewald, *From Kant to Hilbert*, pp. 920–2.

"Foundations of a General Theory of Manifolds: A Mathematico-Philosophical Investigation into the Theory of the Infinite" (1883).

In William Ewald, ed., *From Kant to Hilbert: A Source Book in the Foundations of Mathematics.* Vol. II (Oxford: Clarendon Press, 1996), pp. 381–408.

Cantor, Georg, Richard Dedekind, and David Hilbert. "Cantor's Late Correspondence with Dedekind and Hilbert." In Ewald, *From Kant to Hilbert*, pp. 923–40.

Carlson, Matthew. "Logic and the Structure of the Web of Belief." *Journal for the History of Analytical Philosophy* 3:5 (2015): https://jhaponline.org/jhap/article/view/28.

Carnap, Rudolf. *Foundations of Logic and Mathematics* (Chicago, IL: University of Chicago Press, 1939).

Meaning and Necessity: A Study in Semantics and Modal Logic, 2nd enlarged edn (Chicago, IL: University of Chicago Press, 1956).

"Empiricism, Semantics, and Ontology." In *Meaning and Necessity*, pp. 205–21.

Logical Syntax of Language, trans. Amethe Smeaton (Paterson, NJ: Littlefield, Adams, and Company, 1959).

Logical Foundations of Probability, 2nd edn (Chicago, IL: University of Chicago Press, 1962).

"Intellectual Autobiography." In Schilpp, P. A. *The Philosophy of Rudolf Carnap* (La Salle, IL: Open Court, 1963), pp. 3–84.

Philosophy and Logical Syntax (London: Thoemmes Press, 1996) [originally published 1935].

Church, Alonzo. "Comparison of Russell's Resolution of the Semantical Antinomies with That of Tarski." *Journal of Symbolic Logic* 41: 4 (December 1976), pp. 747–60.

Clark, Peter, Michael Hallett, and David DeVidi, eds. *Vintage Enthusiasms: Essays in Honour of J. L. Bell* (Glasgow ePrints Service: http://eprints.gla.ac.uk/3810, 2008). Accessed 8 September 2018.

Dauben, Joseph Warren. *Georg Cantor: His Mathematics and Philosophy of the Infinite* (Princeton, NJ: Princeton University Press, 1979).

Devlin, Keith. *The Joy of Sets: Fundamentals of Contemporary Set Theory.* 2nd edn (New York, NY: Springer, 1993).

Dreben, Burton. "In Mediis Rebus." *Inquiry* 37 (1994), pp. 441–7.

"Quine." In Barrett and Gibson, *Perspectives on Quine*, pp. 82–95.

Ebbinghaus, Heinz-Dieter. *Ernst Zermelo: An Approach to His Life and Work* (New York, NY: Springer, 2007).

Enderton, Herbert B. *Elements of Set Theory* (San Diego, CA: Academic Press, 1977).

Ewald, William, ed. *From Kant to Hilbert: A Source Book in the Foundations of Mathematics*, vol. II (Oxford: Clarendon Press, 1996).

Feferman, Solomon. "Arithmetization of Meta-Mathematics in a General Setting." *Fundamenta Mathematicae* 69 (1960), pp. 35–92.

Ferreirós, José. *Labyrinth of Thought: A History of Set Theory and Its Role in Modern Mathematics* (Boston, MA: Birkhäuser, 1999).

Floyd, Juliet and Sanford Shieh, eds. *Future Pasts: The Analytic Tradition in Twentieth-Century Philosophy* (New York, NY: Oxford University Press, 2001).

Forster, Thomas. "The Status of the Axiom of Choice in Set Theory with a Universal Set." *Journal of Symbolic Logic* 50:3 (1985), pp. 701–7.

Set Theory with a Universal Set: Exploring an Untyped Universe (Oxford: Clarendon Press, 1992).

"The Iterative Conception of Set." *Review of Symbolic Logic* 1:1 (2008), pp. 97–110.

Fraenkel, A. A., Y. Bar-Hillel, and A. Levy. *Foundations of Set Theory*, 2nd rev. edn (Amsterdam: North-Holland, 1973).

Frege, Gottlob. *The Foundations of Arithmetic: A Logico-Mathematical Enquiry into the Concept of Number*, trans. J. L. Austin, 2nd rev. edn (Evanston, IL: Northwestern University Press, 1968 [1884]).

Grundgesetze der Arithemetik I/II (New York, NY: Georg Olms, 1998 [1893, 1903]).

"Letter to Russell" (1902). In van Heijenoort, *From Frege to Gödel*, pp. 127–8.

Garciadiego, Alejandro. *Bertrand Russell and the Origins of the Set-Theoretic "Paradoxes"* (Boston, MA: Birkhäuser, 1992).

Gentzen, Gerhard. *The Collected Papers of Gerhard Gentzen*. Ed. M. E. Szabo (Amsterdam: North-Holland, 1969).

"The Consistency of Elementary Number Theory." In *Collected Papers*, pp. 132–213.

"The Consistency of the Simple Theory of Types." In *Collected Papers*, pp. 214–22.

Gibson, Roger. *The Philosophy of W. V. Quine: An Expository Essay* (Tampa, FL: University of South Florida Press, 1982).

Enlightened Empiricism: An Examination of W. V. Quine's Theory of Knowledge (Tampa, FL: University of South Florida Press, 1992).

Gödel, Kurt. *Kurt Godel: Collected Works*, 2 vols., Eds. Solomon Feferman, John W. Dawson, Jr., Stephen C. Kleene, Gregory H. Moore, Robert M. Solovay and Jean van Heijenoort (New York, NY: Oxford University Press, 1986, 1990).

"On Formally Undecidable Propositions of Principia Mathematica and Related Systems I." In *Kurt Gödel: Collected Works*, vol. I, pp. 145–95.

"Russell's Mathematical Logic." In Schilpp, *The Philosophy of Bertrand Russell*, pp. 123–53.

"The Consistency of the Axiom of Choice and of the Generalized Continuum Hypothesis with the Axioms of Set Theory." In *Kurt Gödel: Collected Works*, vol. II, pp. 33–101.

"What Is Cantor's Continuum Problem?" In Benacerraf and Putnam, *Philosophy of Mathematics*, pp. 470–85.

Griffin, Nicholas. *Russell's Idealist Apprenticeship* (Oxford: Clarendon Press, 1991).

Gustaffson, Martin. "Quine's Conception of Explication – And Why It Isn't Carnap's." In Gilbert Harman and Ernie Lepore, eds, A Companion to W.V.O. Quine (Malden, MA: Wiley-Blackwell, 2014), pp. 508–25.

Hager, Paul. "Russell's Method of Analysis." In Nicholas Griffin, eds, *The Cambridge Companion to Bertrand Russell* (New York, NY: Cambridge University Press, 2003), pp. 310–31.

Hahn, Edwin and Paul A. Schilpp, eds. *The Philosophy of W. V. Quine*, expanded edn (Chicago, IL: Open Court, 1998).

Hallett, Michael. *Cantorian Set Theory and Limitation of Size* (Oxford: Clarendon Press, 1984).

Hart, W. D. "On Non-Well-Founded Sets." *Crítica* XXIV:72 (1992), pp. 3–20.

The Evolution of Logic (New York, NY: Cambridge University Press, 2010).

Hilbert, David and Paul Bernays. *Grundlagen der Mathematik*, 2 vols., 2nd edn (New York, NY: Springer, 1968/1970).

Hintikka, Jaakko, ed. *Essays on the Development of the Foundations of Mathematics* (Dordrecht: Kluwer Academic Publishers, 1995).

Holmes, Randall. "The Set-Theoretical Program of Quine Succeeded, but Nobody Noticed." *Modern Logic*4 (1994), pp. 1–47.

Elementary Set Theory with a Universal Set. Cahiers du Centre de Logique, vol. 10 (Louvain-la-Neuve: Academia, 1998).

Hrbacek, Karel and Thomas Jech. *Introduction to Set Theory*, 3rd rev. and expanded edn (New York, NY: Marcel Dekker, 1999).

Hylton, Peter. "Analyticity and the Indeterminacy of Translation." *Synthese* 52:2 (1982), pp. 173–5.

Russell, Idealism, and the Emergence of Analytic Philosophy (Oxford: Clarendon Press, 1990).

"Quine on Reference and Ontology." In *The Cambridge Companion to Quine*, ed. Roger Gibson (New York, NY: Cambridge University Press, 2004), pp. 115–50.

Propositions, Functions, and Analysis: Selected Essays on Russell's Philosophy (Oxford: Clarendon Press, 2005).

"Beginning with Analysis." In *Propositions, Functions, and Analysis*, pp. 30–48.

"Frege and Russell." In *Propositions, Functions, and Analysis*, pp. 153–84.

"Logic in Russell's Logicism." In *Propositions, Functions, and Analysis*, pp. 49–82.

"Russell's Substitutional Theory." In *Propositions, Functions, and Analysis*, pp. 83–107.

"The Nature of the Proposition and the Revolt Against Idealism." In *Propositions, Functions, and Analysis*, pp. 9–29.

"The Theory of Descriptions." In *Propositions, Functions, and Analysis*, pp. 189–94.

"The Vicious Circle Principle." In *Propositions, Functions, and Analysis*, pp. 108–14.

Quine (New York, NY: Routledge, 2007).

"Quine's Naturalism Revisited." In Gilbert Harman and Ernie Lepore, eds. *A Companion to W.V.O. Quine* (Malden, MA: Wiley-Blackwell, 2014), pp. 148–62.

Jech, Thomas. *The Axiom of Choice* (Mineola, NY: Dover, 1973).

ed. *Axiomatic Set Theory*, vol. II (Providence, RI: American Mathematical Society, 1974).

Set Theory, 3rd edn (New York, NY: Springer, 2003).

"About the Axiom of Choice." In Barwise, *Handbook of Mathematical Logic*, pp. 345–70.

Jensen, Ronald. "On the Consistency of a Slight (?) Modification of Quine's NF," *Synthese* 19 (1969), pp. 250–63.

Kanamori, Akahiro. "The Mathematical Development of Set Theory from Cantor to Cohen." *The Bulletin of Symbolic Logic* 2:1 (March 1996), pp. 1–71.

The Higher Infinite, 2nd edn (New York, NY: Springer, 2003).

Kleinberg, Eugene M. *Infinitary Combinatorics and the Axiom of Determinateness*. Lecture Notes in Mathematics 612 (New York, NY: Springer).

Kline, Morris, *Mathematical Thought from Ancient to Modern Times* (New York, NY: Oxford University Press, 1972).

Kreisel, George. "On the Mathematical Significance of Consistency Proofs." *Journal of Symbolic Logic* 23:2 (1958), pp. 155–82.

Kunen, Kenneth. *Set Theory: An Introduction to Independence Proofs* (New York, NY: North-Holland, 1980).

Lindenbaum, Adolf and Alfred Tarski. "Communication sur les Recherches de la Théorie des Ensembles" (1926). In Tarski, *Collected Papers*, pp. 171–204.

Linnebo, Øystein. "Sets, Properties, and Unrestricted Quantification." In Rayo and Uzquiano, *Absolute Generality*, pp. 149–78.

MacClane, Saunders. "Locally Small Categories and the Foundations of Set Theory." In *Infinitistic Methods: Proceedings of the Symposium on Foundations of Mathematics* (New York, NY: Pergamon Press, 1961), pp. 25–43.

MacLane, Saunders. "Categorical Algebra and Set-Theoretic Foundations." *Proceedings of Symposia in Pure Mathematics* XIII:pt. I (1971), pp. 231–40.

Maddy, Penelope. *Naturalism in Mathematics* (Oxford: Clarendon Press, 1997).

Defending the Axioms: On the Philosophical Foundations of Set Theory (New York, NY: Oxford University Press, 2011).

Martin, Donald A. "Review of *Set Theory and Its Logic*." *Journal of Philosophy* 67:4 (1970), pp. 111–14.

Mendelson, Elliott. *Introduction to Mathematical Logic* (Boca Raton, FL: CRC Press, 1997).

Mirimanoff, Dimitri. "Les Antinomies de Russell et de Burali-Forti et le Problème Fondamental de la Théorie des Ensembles." *L'Enseignement Mathématique* 19 (1917), pp. 37–52.

"Remarques sur la Théorie des Ensembles." *L'Enseignement Mathématique* 19 (1917), pp. 208–17.

Moore, G. E. *G. E. Moore: The Early Essays*. Ed. Tom Regan (Philadelphia, PA: Temple University Press, 1986), pp. 59–80.

"The Nature of Judgment." In *G. E. Moore: The Early Essays*, pp. 59–80.

Moore, Gregory H. "The Origins of Zermelo's Axiomatization of Set Theory." *Journal of Philosophical Logic* 7 (1978), pp. 307–29.

Zermelo's Axiom of Choice: Its Origins, Development, and Influence (New York, NY: Springer, 1982).

"The Origins of Russell's Paradox: Russell, Couturat, and the Antinomy of Infinite Number." In Jaakko Hintikka, eds, *Essays on the Development of the Foundations of Mathematics* (Dordrecht: Kluwer Academic Publishers, 1995), pp. 215–39.

Moore, Gregory H. and Alejandro Garciadiego. "Burali-Forti's Paradox: A Reappraisal of Its Origins." *Historia Mathematica* 8 (1981), pp. 319–50.

Morris, Sean. "Quine, Russell, and Naturalism: From a Logical Point of View." *Journal of the History of Philosophy* 53:1 (2015), pp. 133–55.

"The Significance of Quine's New Foundations for the Philosophy of Set Theory." *The Monist* 100:2 (2017), 167–79.

Moschovakis, Yiannis. *Descriptive Set Theory* (New York, NY: North-Holland, 1980), Ch. 6.

Notes on Set Theory. 2nd edn (New York, NY: Springer, 2006).

Mycielski, Jan and Stanislaw Swierczkowski, "On the Lebesgue Measureability and the Axiom of Determinateness." *Fundamenta Mathematicae* 54 (1964), pp. 67–71.

Parsons, Charles. "What Is the Iterative Conception of Set?" In Benacerraf and Putnam, *Philosophy of Mathematics*, pp. 503–29.

Peirce, Charles. *Collected Papers*. Vol. II (Cambridge, MA: Harvard University Press, 1932.

Potter, Michael. "Iterative Set Theory." *The Philosophical Quarterly* 43:171 (1993), pp. 178–93.

Set Theory and Its Philosophy (New York, NY: Oxford University Press, 2004).

Quine, W. V. *A System of Logistic* (Cambridge, MA: Harvard University Press, 1934).

"New Foundations for Mathematical Logic," *American Mathematical Monthly* 44:2 (1937), pp. 70–80.

"On Cantor's Theorem." *Journal of Symbolic Logic* 2:3 (September 1937), pp. 120–4.

"Unification of Universes in Set Theory." *Journal of Symbolic Logic* 21:3 (1956), pp. 267–79.

Word and Object (Cambridge, MA: MIT Press, 1960).

Ontological Relativity and Other Essays (New York, NY: Columbia University Press, 1969).

"Epistemology Naturalized." In *Ontological Relativity and Other Essays*, pp. 69–90.

Set Theory and Its Logic, rev. edn (Cambridge, MA: Harvard University Press, 1969).
"Reply to Martin." *Journal of Philosophy* 67:8 (1970), pp. 247–8.
Methods of Logic, 3rd edn (Chicago, IL: Holt Rinehart, and Winston, 1972).
The Ways of Paradox and Other Essays, rev. and enlarged ed. (Cambridge, MA: Harvard University Press, 1976).
"Foundations of Mathematics" (1965). In *The Ways of Paradox and Other Essays*, pp. 22–32.
"Mr. Strawson on Logical Theory" (1953). In *The Ways of Paradox and Other Essays*, pp. 137–57.
"Necessary Truth" (1963). In *The Ways of Paradox and Other Essays*, pp. 68–76.
"Posits and Reality" (1955). In *The Ways of Paradox and Other Essays*, pp. 246–54.
From a Logical Point of View: Nine Logico-Philosophical Essays, 2nd rev. edn (Cambridge, MA: Harvard University Press, 1980).
"New Foundations for Mathematical Logic" with supplementary remarks. In *From a Logical Point of View*, pp. 80–101.
"On What There Is" (1948). In *From a Logical Point of View*, pp. 1–19.
"Two Dogmas of Empiricism" (1951). In *From a Logical Point of View*, pp. 20–46.
Theories and Things (Cambridge, MA: Harvard University Press, 1981).
"On the Individuation of Attributes" (1975). In *Theories and Things*, pp. 100–12.
Methods of Logic, 4th edn (Cambridge, MA: Harvard University Press, 1982).
The Time of My Life (Cambridge, MA: MIT Press, 1985).
The Logic of Sequences: A Generalization of Principia Mathematica (1932) (New York, NY: Garland, 1990).
Selected Logic Papers, enlarged edn (Cambridge, MA: Harvard University Press, 1995).
"A Bibliography of the Publications of W. V. Quine." In Hahn and Schilpp, *Philosophy of W. V. Quine*, pp. 743–64.
"Autobiography of W. V. Quine." In Hahn and Schilpp, *Philosophy of W. V. Quine*, pp. 1–46, 729–41.
"Confessions of a Confirmed Extensionalist." In Floyd and Shieh, *Future Pasts*, pp. 215–21.
"Set-Theoretic Foundations for Logic" (1936). In *Selected Logic Papers*, pp. 83–99.
"The Inception of New Foundations" (1987). In *Selected Logic Papers*, pp. 286–9.
"Whitehead and the Rise of Modern Logic" (1941). In *Selected Logic Papers*, pp. 3–36.
"Carnap and Logical Truth," Reprinted in W. V. Quine, *The Ways of Paradox and Other Essays*, rev. and enlarged edn (Cambridge, MA: Harvard University Press, 1976), pp. 107–32.

"Concepts of Negative Degree." In *Proceedings of the National Academy of Sciences of the United States of America* 22:1 (1936), pp. 40–5.

From Stimulus to Science (Cambridge, MA: Harvard University Press, 1995).

"On the Axiom of Reducibility." *Mind* 45:180 (1936), pp. 488–500.

"On Derivability," *Journal of Symbolic Logic* 2:3 (1937), pp. 113–19.

"On the Theory of Types," *Journal of Symbolic Logic* 3:4 (1938), pp. 125–39.

"A Theory of Classes Presupposing No Canons of Type." In *Proceedings of the National Academy of Sciences* 22 (1936), 320–26.

"Truth by Convention." In *The Ways of Paradox and Other Essays*, rev. and enlarged edn (Cambridge, MA: Harvard University Press, 1976), pp. 77–106.

"Ontological Relativity." In *Ontological Relativity and Other Essays*, pp. 26–48.

The Roots of Reference (La Salle, IL: Open Court, 1972), pp. 102–3.

"Set-Theoretic Foundations for Logic," *Journal of Symbolic Logic* 1 (1936), pp. 45–57.

"The Ways of Paradox" (1962). In *The Ways of Paradox and Other Essays*, pp. 1–18.

"Two Dogmas in Retrospect." Reprinted in W. V. Quine, *Confessions of a Confirmed Extensionalist and Other Essays.* Eds. Dagfinn Føllesdal and Douglas B. Quine (Cambridge, MA: Harvard University Press, 2008), pp. 390–401.

"Kurt Gödel" (1979). In *Theories and Things*, pp. 143–7.

Philosophy of Logic, 2nd edn (Cambridge, MA: Harvard University Press, 1986).

"The Scope and Language of Science." Reprinted in W. V. Quine, *The Ways of Paradox and Other Essays*, rev. and enlarged edn (Cambridge, MA: Harvard University Press, 1976), pp. 228–45.

"Things and Their Place in Theories." In W. V. Quine, *Theories and Things* (Cambridge, MA: Harvard University Press, 1981), pp. 1–23.

Pursuit of Truth, rev. ed. (Cambridge, MA: Harvard University Press, 1992).

"Reply to Jules Vuillemen." In Hahn and Schilpp, *The Philosophy of W. V. Quine*, pp. 619–20.

Quine, W. V. and Joseph Ullian. *The Web of Belief*, 2nd edn (New York, NY: Random House, 1978).

Ramsey, F. P. "The Foundations of Mathematics." *Proceedings of the London Mathematical Society*, s2–25:1 (1926), pp. 338–84.

The Foundations of Mathematics and Other Logical Essays. Ed. R. B. Braithewaite (New York, NY: Humanities Press, 1950).

"The Foundations of Mathematics" (1925). In *The Foundations of Mathematics*, pp. 1–61.

Rang, B. and W. Thomas, "Zermelo's Discovery of the 'Russell Paradox.'" *Historia Mathematica* 8 (1981), pp. 15–22.

Rayo, Agustin and Gabriel Uzquiano, eds. *Absolute Generality* (New York, NY: Oxford University Press, 2006).

Reck, Erich. "Carnapian Explication: A Case Study and a Critique." In Pierre Wagner, ed., *Carnap's Ideal of Explication and Naturalism* (New York, NY: Palgrave-Macmillan, 2012), pp. 96–116.

Resnick, Michael D. "On the Philosophical Significance of Consistency Proofs." *Journal of Philosophical Logic* 3:1/2 (1974), pp. 133–47.

Robinson, Abraham. "Non-Standard Analysis." *Selected Papers*, vol. 2. Eds, H. J. Keisler et al. (New Haven, CT: Yale University Press, 1979), pp. 3–11.

Ricketts, Thomas. "Languages and Calculi." In Gary L. Hardcastle and Alan W. Richardson, eds, *Logical Empiricism in North America* (Minneapolis, MN: University of Minnesota Press, 2003), pp. 257–80.

Rieger, Adam. "An Argument for Finsler-Aczel Set Theory." *Mind* 109:434 (2000), pp. 241–53.

"Paradox, ZF, and the Axiom of Foundation." In Clark, Hallett, and DeVidi, *Vintage Enthusiasms*, pp. 1–22.

"Paradox, ZF, and the axiom of foundation." In Peter Clark, Michael Hallett, and David DeVidi, eds, *Logic, Mathematics, Philosophy, Vintage Enthusiasms: Essays in Honour of John L. Bell.* Springer (2011), pp. 171–187. ISBN 9789400702134 (doi:10.1007/978-94-007-0214-1_9).

Rosser, J. Barkley. "On the Consistency of Quine's New Foundations for Mathematical Logic." *Journal of Symbolic Logic* 4:1 (1939), pp. 15–24.

Logic for Mathematicians (New York, NY: McGraw-Hill, 1953).

"Review of Specker's 'The Axiom of Choice in Quine's New Foundations for Mathematical Logic.'" *Journal of Symbolic Logic* 19:2 (1954), pp. 127–8.

Logic for Mathematicians, 2nd edn with additional appendices (Mineola, NY: Dover Publications, 1978).

Russell, Bertrand. *Introduction to Mathematical Philosophy* (London: George Allen and Unwin, 1919).

The ABC of Relativity (New York, NY: Harper, 1925).

Philosophy (New York, NY: W. W. Norton, 1927).

Sceptical Essays (New York, NY: W. W. Norton, 1928).

Marriage and Morals (New York, NY: Horace Liveright, 1929).

Our Knowledge of the External World (New York, NY: W. W. Norton, 1929).

Principles of Mathematics, 2nd edn (London: George Allen and Unwin, 1937 [first edition 1903]).

Logic and Knowledge: Essays 1901–1950. Ed. Robert C. Marsh (London: George Allen and Unwin, 1956).

"On Denoting" (1905). In *Logic and Knowledge*, pp. 39–57.

Essays in Analysis. Ed. Douglas Lackey (London: George Allen and Unwin, 1973).

"On Some Difficulties in the Theory of Transfinite Ordinals and Order Types" (1905). In *Essays in Analysis*, pp. 135–64.

"The Regressive Method of Discovering the Premises of Mathematics." In *Essays in Analysis*, pp. 272–83.

The Collected Papers of Bertrand Russell, vol. 2. Eds. Nicholas Griffin and Albert C. Lewis (New York, NY: Routledge, 1990).

My Philosophical Development (New York, NY: Routledge, 1993 [1958]).
"Mathematical Logic as Based on the Theory of Types" (1908). In van Heijenoort, *From Frege to Gödel*, pp. 150–82.
The Collected Papers of Bertrand Russell, vol. 3. Ed. Gregory H. Moore (New York, NY: Routledge, 1993).
"On Some Difficulties of Continuous Quantity" (1896). In *The Collected Papers*, pp. 44–58.
"Review of Essai critique sur la hypothèse des atomes dans la science contemporaine" (1896). In *The Collected Papers*, pp. 35–43.
Schilpp, Paul Arthur, ed. *The Philosophy of Bertrand Russell* (La Salle, IL: Open Court Press, 1971).
Scott, Dana. "Axiomatizing Set Theory." In Jech, *Axiomatic Set Theory*, vol. II, pp. 207–14.
Shoenfield, Joseph. *Mathematical Logic* (Natick, MA: A. K. Peters, 1967).
"Axioms of Set Theory." In Barwise, *Handbook of Mathematical Logic*, pp. 321–44.
Sierpinski, Waclaw. "L'hypothèse généralisée du continu et l'axiom du choix." *Fundamenta Mathematica* 34 (1947), pp. 1–5.
Specker, Ernst P. "The Axiom of Choice in Quine's New Foundations for Mathematical Logic." *Proceedings of the National Academy of Sciences* 39:9 (September 1953), pp. 972–5.
Szabo, M. E., ed. *The Collected Papers of Gerhard Gentzen* (Amsterdam: North-Holland, 1969).
Tait, William. "Frege Versus Cantor and Dedekind: On the Concept of Number." In *The Provenance of Pure Reason: Essays in the Philosophy of Mathematics and Its History* (New York, NY: Oxford University Press, 2005), pp. 212–51.
The Provenance of Pure Reason: Essays in the Philosophy of Mathematics and Its History (New York, NY: Oxford University Press, 2005).
Tarski, Alfred. *Collected Papers*, vol. I, 1921–34. Eds Steven Givant and Ralph N. McKenzie (Boston, MA: Birkhäuser, 1986).
Ullian, Joseph. "Quine and the Field of Mathematical Logic." In Hahn and Schilpp, *Philosophy of W. V. Quine*, pp. 569–89.
van Heijenoort, Jean, ed. *From Frege to Gödel: A Source Book in Mathematical Logic, 1879–1931* (Cambridge, MA: Harvard University Press, 1967).
von Neumann, John. "An Axiomatization of Set Theory" (1925). In van Heijenoort, *From Frege to Gödel*, pp. 393–413.
John von Neumann: Collected Works, vol. I. Ed. Abraham H. Taub (New York, NY: Pergamon Press, 1961).
"Über eine Widerspruchfreiheitsfrage in der axiomatischen Mengenlehre" (1929). In *Collected Works*, pp. 494–508.
Wagon, Stan. *The Banach-Tarski Paradox* (Cambridge: Cambridge University Press, 2005).
Wang, Hao. "The Concept of Set." In Benacerraf and Putnam, eds., *Philosophy of Mathematics*, pp. 530–70.

Weir, Alan. "Is It Too Much to Ask, to Ask for Everything?" In Rayo and Uzquiano, *Absolute Generality*, pp. 333–68.

Whitehead, Alfred North and Bertrand Russell. *Principia Mathematica*, 3 vols. 2nd edn (Cambridge: Cambridge University Press, 1925, 1927 [first edn 1910, 1912, 1913]).

Wittgenstein, Ludwig. *Tractatus Logico-Philosophicus* (New York, NY: Routledge, 1990 [1922]).

Cambridge Letters: Correspondence with Russell, Keynes, Moore, Ramsey, and Sraffa. eds. Brian McGuinnes and Georg Henrik von Wright (Malden, MA: Blackwell Publishers, 1997).

Zermelo, Ernst. "A New Proof of the Possibility of a Well-Ordering" (1908). In van Heijenoort, *From Frege to Gödel*, pp. 183–98.

"Investigations into the Foundations of Set Theory I" (1908). In van Heijenoort, *From Frege to Gödel*, pp. 199–215.

"On Boundary Numbers and Domains of Sets: New Investigations in the Foundations of Set Theory" (1930). In Ewald, *From Kant to Hilbert*, pp. 1219–33.

"Proof That Every Set Can Be Well-Ordered" (1904). In van Heijenoort, *From Frege to Gödel*, pp. 139–41.

Index